Electronic Circuit Guidebook

Volume 5:
Digital Electronics

Electronic Circuit Guidebook

Volume 5:
Digital Electronics

Written By
Joseph J. Carr

PROMPT®
PUBLICATIONS

PROMPT© Publications is an imprint of Howard W. Sams & Company, A Bell Atlantic Company, 2647 Waterfront Parkway, E. Dr., Indianapolis, IN 46214-2041.

International Standard Book Number: 0-7906-1129-5
Library of Congress Catalog Card Number: 97-65787

Acquisitions Editor: Loretta Yates
Editor: Natalie F. Harris
Assistant Editors: Pat Brady, J.B. Hall
Typesetting: Natalie Harris
Indexing: Natalie Harris
Cover Design: Christy Pierce
Graphics: Joe Carr
Graphics Conversion: Terry Varvel

PRINTED IN THE UNITED STATES OF AMERICA

9 8 7 6 5 4 3 2 1

Contents

Preface

Have you noticed the "Digital Revolution" in electronics? If you haven't, then you must have been in suspended animation for the past twenty years or so, and certainly since the early 1980s. Nearly everything electronic is "coming up digital."

Digital electronics includes personal computers, large computers, hand-held calculators, industrial instruments, test equipment, household appliances, and a host of other devices. Even traditionally analog devices such as television sets and radio receivers contain a large amount of digital circuitry. Some products are made better using the digital implementation, while still others are made either possible or affordable by the use of digital circuitry.

This book is about the basics of digital electronics. It is not a computer book, but rather it is about the fundamental circuits that make up not only computers, but all digital products. It joins a series of other titles in the Electronic Circuit Guidebook series in providing you with encapsulated practical information about specific areas of electronics technology.

Unlike certain other current books on digital circuits, this book will not spend any time on obsolete digital devices. At one time, diode-transistor-logic (DTL) and resistor-transistor-logic (RTL) were current. Even at relatively late dates covering these families was reasonable because repair technicians occasionally encountered them. However, those devices have faded so far into the past that it no longer makes sense to cover them at all.

This text is not bound to any particular logic family (we examine the modern versions), but rather it discusses the logic elements from a functional point of view. That is, we will look at logic gates, flip-flops, counters and the ways they are combined to make practical circuits.

—Joseph J. Carr, MSEE

Chapter 1

Introduction to Digital Electronics

The origins of "digital electronics" as a separate field are difficult to pin down historically. One might argue, for example, that the first "on/off" switch was "digital" (although the argument would be a bit specious).

So what are "digital" circuits? A digital circuit is one that has only two possible states; that is, it is binary in nature. These states are represented by the binary digits "0" and "1." Because binary digits ("bits") are used, digital electronic circuits can be described by the base 2 number system.

At one time, digital electronics were limited to making calculations (computing) and certain simple control functions. Although no one can successfully argue that analog electronics are down and dead, digital electronics have taken over many functions that were once analog (filtering, control systems, etc.). Tremendous growth has also occurred in areas of electronics that have traditionally seen heavy digital involvement. This growth accelerated to light-speed proportions by the invention of the microprocessor chip in the 1970s. Today, microprocessors are available with clock speeds that sound like radar frequencies, and the progress continues.

Revolution in Work Practices

The digital revolution is truly a new industrial revolution, and has had about the same impact on human working habits as the original industrial revolution of the 18th and 19th centuries. Automobiles, for example, were once very analog in their

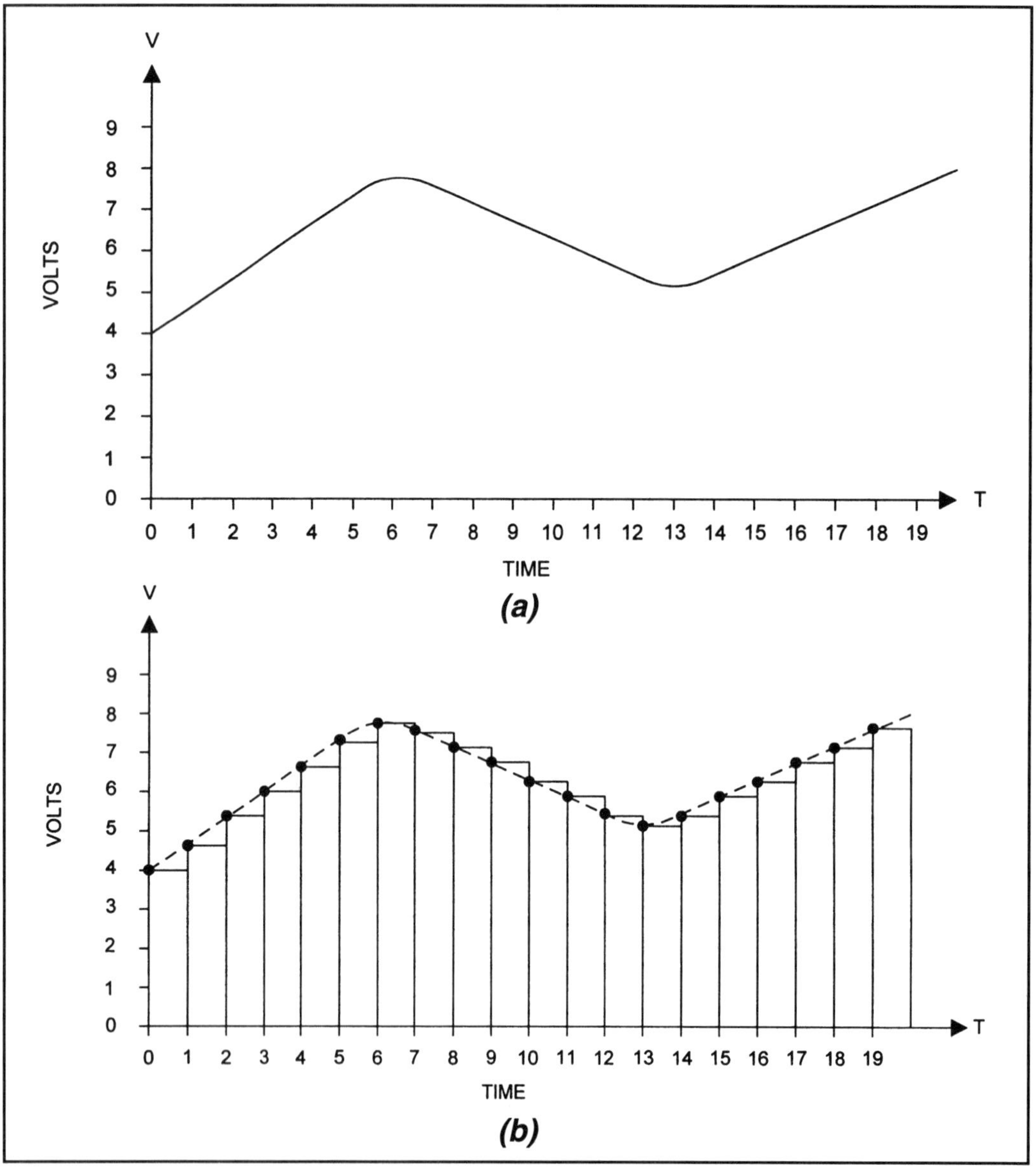

***Figure 1-1**. a. Analog signal; b. sampled signal.*

electrical systems. But today, chip makers are selling automobile manufacturers nearly 3,000,000 chips a day. One of my own cars has 43 processors on board, according to the owner's manual. Car makers are now speaking the same language of data busses and I/O ports that computer, information technology and military users speak. The car of the near future will have a fiber optic data bus on board that controls every facet of the vehicle's operation. Digital electronics will also make new functions possible, such as global positioning systems for finding your location and moving digital maps for figuring out how to get to where you want to go.

Analog vs. Digital Signals

There are three basic forms of circuitry used in electronics (at least according to one way of reckoning): *analog*, *sampled*, and *digital.*

Analog signals (*Figure 1-1a*) are continuous in both time and amplitude. The analog signal can take on any value between the limits. For example, if the limits are 0 to 10 volts, then the analog signal V(t) can take on any value between those limits. For example, a value can be 5.0 volts, 5.5 volts, or 7.54232233 volts. The precision of your knowledge of the voltage is limited solely by your ability to measure it (most practical meters won't measure 7.54232233 volts, but will probably round off at 7.542 or so).

A sampled signal (*Figure 1-1b*) is discrete in time but continuous in amplitude. The sampling takes place at specific time intervals, so the time element is discrete (limited to those points at which a sample is taken). The amplitude, on the other hand, can take on any value between limits.

The digital signal is discrete in both time and amplitude, so is best represented by a table of values (*Table 1-1*). A digital representation of an analog or sampled signal divides the amplitude range into a specified number of different discrete steps. For example, an 8-bit digital system breaks the range into 256 different steps, or 255 intervals if zero is taken into account. If the range is 0 to 10 volts, then each step represents a change of 39 mV.

Note that all three types of signal can denote the same information, but only the digital signal can be used in a computer. The other two forms must be first converted to binary format so that the computer can understand them. In data acquisition systems, it is common to find sampled signals that have been digitized. Indeed, the fact that the *analog-to-digital converter* (ADC) used to convert the analog signal to digital form has a specified *conversion time*, Tc (time required to make the conversion), makes the signals inherently sampled. Thus, the values stored in the computer are, in essence, sampled at Tc intervals.

Time	Decimal	Binary
0	16	10000
1	18	10010
2	22	10110
3	24	11000
4	26	11010
5	29	11101
6	30	11110
7	29	11101
8	28	11100
9	27	11011
10	25	11001
11	23	10111
12	22	10110
13	20	10100
14	22	10110
15	25	11001
16	26	10010
17	27	11011
18	28	11100
19	31	11111

***Table 1-1**. A digital signal is discrete in both time and amplitude.*

Digital Electronics Circuit Elements

Digital electronics circuits are based on a few elementary devices called *gates*. The circuit symbols for these elements are shown in *Figure 1-2*, and their function will be described in greater detail in Chapter 5. For now, however, we will just state that a logic gate is a device that will pass or not pass binary signals according to a limited set of fixed rules.

Logic Levels

Digital logic devices operate on binary signals — signals that have only two states. These represent the binary (base 2) digits 0 and 1. These states are sometimes called by the designations below:

	Digit	A.K.A.
0	Logical 0	LOW
1	Logical 1	HIGH

In digital electronic circuits, there are a number of different schemes for representing 0 and 1. I won't bother to cover some of the older systems, except for the RS-232C serial port standard used on computers.

Perhaps the most common scheme is that originally used on the transistor-transistor-logic (TTL) family of devices. These devices are operated from 0VDC and +5VDC power supplies. The logic levels were represented by 0 to 0.8 volts for LOW (0) and > +2.4 volts for HIGH (1). The CMOS family is usually operated at the TTL levels,

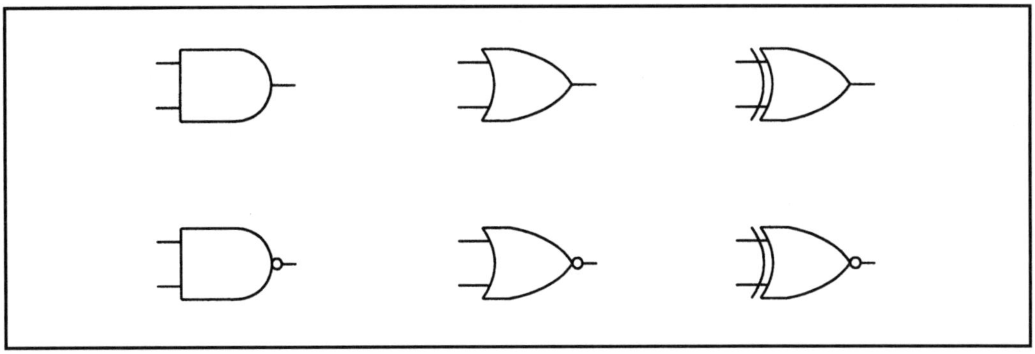

Figure 1-2. Logic gate circuit symbols

but can be used at any voltage up to as much as -18 volts and +18 volts for LOW and HIGH. The RS-232C serial communications port standard uses -5 to -15 volts for the HIGH (1) and +5 to +15 volts for the LOW (0).

Before covering the logic gates we will examine digital number systems (Chapter 2), digital coding schemes (Chapter 3), and the digital logic families (Chapter 4).

Chapter 2

Number Systems for Digital Electronics

One of the differences between analog and digital electronics is the way in which you handle the numbers. In normal life, we use the decimal numbers system, which is based on the number 10. It is also called a *base 10 numbers system* because of this fact. Digital electronics, on the other hand, is binary — it has only two states, 0 and 1; so the arithmetic of digital electronics is based on binary or base 2 numbers. A kind of shorthand is sometimes used, however, because binary numbers are hard to work with. Base 8 (octal) and base 16 (hexadecimal) numbers are used when discussing digital systems, rather than base 2 (even though the circuits are base 2 oriented). This can be done because 8 and 16 are powers of 2 ($2^3 = 8$ and $2^4 = 16$).

In this chapter, we will look at the base 2, base 8 and base 16 arithmetics. In order to familiarize you with the concepts, however, we are going begin by presenting the same sort of material in the base 10 decimal system. This way, you will have a certain comfort level when we start talking about base 2, base 8 and base 16 systems. Just keep in mind one thing if you get confused: *base 2 is just like base 10 if you're missing nine fingers.*

A Word About Notation

The same digits overlap in all of the number systems that we use. In base 10, for example, the digits are 0-9. In base 8, they are 0-7, and in base 2, 0 and 1. In base 16, the digits 0-9 are used, with A-F used to represent the digits 10 through 15. So, is a number that is written "101" the decimal one-hundred and one, or the binary equivalent of decimal five? Or is it an octal equivalent of decimal 65, or hexadecimal 257? In order to prevent confusion, a notation system is needed. We will use subscripts to denote the base of the number:

N_2 Binary (base 2)
N_8 Octal (base 8)
N_{10} Decimal (base 10)
N_{16} Hexadecimal (base 16)

In some cases, an alternate notation for the base 16 hexadecimal number system is to precede the number with the letter "H." For example, either "10116" or "H101" would denote the hexadecimal equivalent of decimal 257.

Decimal Numbers System (Base 10)

The base 10 ("decimal") number system is the one that we normally use in our "real world." If you passed third grade, then you know the basics of the decimal numbers system. And even if you failed third grade, but are able to figure out whether the cashier gave you the correct change when you bought this book, then you understand decimal numbers. Because we know the decimal so well and its concepts are familiar to us, let's take a look at the decimal system as an example from which to build information on the binary, octal and hexadecimal systems.

All four numbers systems are examples of *weighted numbers systems*. The actual value of a weighted number depends on the *digits* used (such as 0-9) and their positions relative to each other.

The decimal, or base 10, number system uses ten digits: 0, 1, 2, 3, 4, 5, 6, 7, 8 and 9. These digits each represent different quantities, or values, that are already familiar to you.

With only ten digits, however, we can represent only ten different quantities. But instead of creating a different digit — a digit is merely a symbol — to represent each possible quantity from zero to infinity, *positional notation* is used. This means that a digit's position relative to other digits confers added value. For example, the decimal number 682_{10} means:

$$682_{10} = (6 \text{ x } 10^2) + (8 \text{ x } 10^1) + (2 \text{ x } 10^0)$$

— or,

$$682_{10} = (600) + (80) + (2)$$

The base of the number system, 10 in this case, is called the *radix*. In general, all weighted number systems use a digit method multiplied by a radix raised to some power at each position. The general form for all weighted number systems is:

$$D_N R^N + \ldots + D_3 R^3 + D_2 R^2 + D_1 R^1 + D_0 R^0$$

— where the D terms are the digits (0-9 in base 10 systems), and the R term is the radix (10 in base 10 systems). In the notation above, the *most significant digit* (MSD) is the one all the way to the left ($D_N R^N$), while the *least significant digit* (LSD) is the one all the way to the right ($D_0 R^0$).

When we place our example number, 682_{10}, in the general form it is as shown above, $(6 \text{ x } 10^2) + (8 \text{ x } 10^1) + (2 \text{ x } 10^0)$.

Note: Any number raised to the zero power (10^0) is equal to one; so $10^0 = 1$.

We can use the general form to represent either whole or partial quantities. A partial quantity is a fraction. A decimal point must be used to separate the two. The digits to the left of the decimal point represent the whole quantities, while those to the right represent the fractional quantities. For example, 18.23 means:

$$\begin{aligned} 18.23_{10} &= (1 \text{ x } 10^1) + (8 \text{ x } 10^0) + (2 \text{ x } 10^{-1}) + (3 \text{ x } 10^{-2}) \\ &= (10) + (8) + (2 \text{ x } 0.1) + (3 \text{ x } 0.01) \\ &= 10 + 8 + 0.2 + 0.03 = 18.23 \end{aligned}$$

Binary Numbers

The binary number system is base 2 and recognizes only the digits 0 and 1. In the notation of the general form, a weighted binary number would be:

$$D_N2^N + \ldots D_32^3 + D_22^2 + D_12^1 + D_02^0$$

— where the D terms are the digits 0 and 1.

For example, the binary number 01011_2 is the shorthand way of writing:

$$(0 \text{ x } 2^4) + (1 \text{ x } 2^3) + (0 \text{ x } 2^2) + (1 \text{ x } 2^1) + (1 \text{ x } 2^0)$$
$$= (0) + (8) + (0) + (2) + (1) = 11_{10}$$

At this point, we'll illustrate why the subscript notation must be used to let you know which system is being used. The digits "10" do not represent the same quantity in binary and decimal systems:

$$10_2 = (1 \text{ x } 2^1) + (0 \text{ x } 2^0) = 2_{10}$$
$$10_{10} = (1 \text{ x } 10^1) + (0 \text{ x } 10^0) = 10_{10}$$

Obviously, $10_2 \neq 10_{10}$.

The term *bit* in digital electronics is a shorthand way of saying *binary digit*, and it is the smallest possible unit of data. A bit can be either of the binary digits, 0 or 1. A *byte* is an array of eight bits arranged in a positional number representation from 00000000 to 11111111. Some people use the term *nybble* to refer to a four-bit number, but this is not "official."

Octal Numbers

The octal number system is just like the decimal number system if you're missing two fingers. The octal system is base 8, and uses the digits 0, 1, 2, 3, 4, 5, 6 and 7. Using the notation of the general form equation, a weighted octal number would look like this:

$$D_N 8^N + \ldots D_3 8^3 + D_2 8^2 + D_1 8^1 + D_0 8^0$$

— where the D terms are the digits 0 through 7. In the standard notation, then, 362_8 would be:

$$(3 \text{ x } 8^2) + (6 \text{ x } 8^1) + (2 \text{ x } 8^0)$$
$$= (3 \text{ x } 64) + (6 \text{ x } 8) + (2 \text{ x } 1)$$
$$= (192) + (48) + (2) = 242_{10}$$

Hexadecimal Numbers

The hexadecimal number system could be just as simple as the decimal number system, but you'd have to grow an five additional fingers. The hexadecimal system — nicknamed *hex* — is base 16, so it uses the sixteen digits 0 through 15. But because all digits greater than 9 are double digits (10, 11, 12, 13, 14 and 15), and a single symbol is needed to represent each digit, the letters A through F are used to represent the digits 10 through 15. The digits of the hexadecimal number system are, therefore, 0, 1, 2, 3, 4, 5, 6, 7, 8, 9, A, B, C, D, E and F, where:

A = 10
B = 11
C = 12
D = 13
E = 14
F = 15

In the standard notation, a hexadecimal number would look like:

$$D_N 16^N + \ldots D_3 16^3 + D_2 16^2 + D_1 16^1 + D_0 16^0$$

The hex number $7F6D_{16}$ is equal to:

$$7F6D_{16} = (7 \text{ x } 16^3) + (F \text{ x } 16^2) + (6 \text{ x } 16^1) + (D \text{ x } 16^0)$$
$$= (7 \text{ x } 4096) + (15 \text{ x } 256) + (6 \text{ x } 16) + (13 \text{ x } 1)$$
$$= 28{,}672 + 3{,}840 + 96 + 13 = 32{,}621_{10}$$

Number System Conversions

It sometimes becomes necessary to convert from one number system to another. This is sometimes done by service technicians and field engineers because they must correlate the numbers in equipment documentation with what they see in the field (of course, some digital analyzers do this for them). Some hand-held calculators will do the conversion, as will the scientific version of the calculator bundled with a computer's *Windows* operating system. For others, the manual method is as follows:

1. Write down the number being converted.

2. Multiply the most significant digit (left-most) by the radix of the number.

3. Add the result of step number 2 to the next most significant digit (MSD-1).

4. Multiply the result obtained in step number 3 by the radix.

5. Repeat steps 3 and 4 until all of the digits are exhausted.

Use *Table 2-1* as a guide. The procedure above converts any number in any base, R, to the equivalent decimal number. The following procedure does the opposite — it converts a decimal number to an equivalent in base R:

1. Divide the decimal number by the radix of the new number system.

2. The remainder of this step becomes the least significant digit (LSD) of the equivalent number in base R.

3. Divide the quotient from step number 1 by the radix.

4. The remainder from this operation becomes the next digit in the new number (LSD+1).

5. Continue until all of the digits are exhausted.

These procedures allow you to convert into decimal form from radix R, or from decimal into radix R. It is usually less trouble to use both procedures when convert-

Decimal	Hex	Octal	Binary	Decimal	Hex	Octal	Binary
0	0	0	0	16	10	20	10000
1	1	1	1	17	11	21	10001
2	2	2	10	18	12	22	10010
3	3	3	11	19	13	23	10011
4	4	4	100	20	14	24	10100
5	5	5	101	21	15	25	10101
6	6	6	110	22	16	26	10110
7	7	7	111	23	17	27	10111
8	8	10	1000	24	18	30	11000
9	9	11	1001	25	19	31	11001
10	A	12	1010	26	1A	32	11010
11	B	13	1011	27	1B	33	11011
12	C	14	1100	28	1C	34	11100
13	D	15	1101	29	1D	35	11101
14	E	16	1110	30	!E	36	11110
15	F	17	1111	31	1F	37	11111
				32	20	40	100000
				33	21	41	100001

Table 2-1. *Number system conversion guide.*

ing from radix R1 to radix R2, when neither R1 nor R2 is decimal. For example, when converting from hex to octal, convert first from hex to decimal and then from decimal to octal.

Binary Arithmetic

Binary arithmetic is occasionally useful when trying to figure out complex digital circuits. There are five basic functions: *addition*, *subtraction*, *twos complement subtraction*, *multiplication* and *division*.

Addition

The rules governing binary addition are simple (some people think they are simpler than decimal addition):

$0 + 0 = 0$
$1 + 0 = 1$

0 + 1 = 1
1 + 1 = 0 plus carry 1

For example, add the binary numbers 01001_2 and 01110_2:

```
  1
  0 1 0 0 1
+ 0 1 1 1 0
-----------
  1 0 1 1 1
```

Subtraction

The rules for straight binary subtraction are as follows:

0 - 0 = 0
0 - 1 = 1 borrow 1
1 - 0 = 0
1 - 1 = 0

Note, however, that it is difficult to build a digital electronic circuit that will perform straight binary subtraction operations. It is, however, quite trivial to build binary adder circuits. In most digital devices required to perform subtraction, therefore, it is simpler to use the twos complement form of subtraction to trick the circuit into thinking it's adding.

The complement of any binary number is its inverse; that is, the complement of 1 is 0 and the complement of 0 is 1. To form the ones complement of any binary number, we change all of the ones to zeros and all of the zeros to ones:

Digit	**Complement**
0	1
1	0

For example, the ones complement of 100110011 is 01001100. Note that all of the ones became zeros and zeros became ones.

Twos Complement Subtraction

The twos complement of a binary number is formed by adding 1 to the ones complement of the number. For example, to find the two's complement of 1011001_2:

1 0 1 1 0 0 1	Number
0 1 0 0 1 1 0	Ones Complement
+ 1	Add 1
0 1 0 0 1 1 1	Twos Complement

You can subtract a binary subtratend from a binary minuend by adding the twos complement of the subtratend to the minuend. Keep in mind that subtraction can be viewed as adding a negative number, and the twos complement notation is a form of negative binary numbers.

Multiplication

Binary multiplication can be done using the same method as decimal multiplication, or by repetitive addition. Multiplication is, after all, nothing more than addition performed over and over. For example, in decimal: 4 x 5 = 4 + 4 + 4 + 4 + 4 = 20, or five repetitions of the +4 operation.

The same method can be used on binary numbers. The technique is called *repetitive addition*, or the *left shift method.* The technique is:

1. Set the answer initially to zero, or 00000000_2.

2. Write down the multiplication so that the least significant digit of the accumulated answer is aligned with the LSD of the multiplicand.

3. Inspect the multiplier beginning with the LSD.

   a. If it is a 1, then add the multiplicand to the answer
   b. If it is a 0, then add 0 to the answer. Shift the multiplicand one bit to the left.

4. Repeat steps 2 and 3 until finished.

The left shift method seems a little cumbersome. On paper this is true, but in digital electronic circuits, adding and left shifting are very easy to accomplish. In fact, it is difficult to do other operations, but shifting and addition are duck soup.

Logical Operations

Digital circuits perform certain *logical operations* called NOT, OR, AND and exclusive OR (XOR). These basic functions also have inverse variations called NOT OR (NOR), NOT AND (NAND) and NOT XOR (NXOR). In each case, two or more bits are compared, and the logical outcome is determined by the basic function being used.

Again, consider some matters of notation. The NOT function is the *inverting* function. In the NOT function, a 0 becomes 1 and a 1 becomes 0. We represent the NOT function either with the word NOT or as a symbol with a bar over it:

$$\text{NOT A} = \overline{A}$$

In text, the NOT A is sometimes written either *NOT A* or *A bar*.

If the NOT symbol is used in conjunction with other functions, then the function is inverted. The OR function becomes the NOT OR, for example.

In one notation system, the multiplication symbol (x) represents the AND function, and the addition symbol (+) is used to represent the OR function.

Chapter 3

Digital Codes

In this chapter, you will learn the most common codes used to represent binary data. A digital code is a binary representation of data. There is nothing mysterious about codes; they are merely methods for presenting data in formats that a computer can recognize. A code may represent alphabetic, numerical, alphanumeric, control signals, punctuation and other forms of data. Most computers have a keyboard, which translates your keystrokes into digital codes that let the computer know your intentions. The keyboard code contains the numbers, letters and punctuation for the language being used. It also has codes for functions such as *shift*, *carriage return* and *line feed*.

Binary numbers are a species of "code" used to represent numerical quantities; but ordinarily, a *binary code* for a numeral does not correspond to the base 2 binary number with the same bit structure. For example, in the standard code used for keyboards, the numerical keyboard character "7" is 00110111, while the binary number that represents the decimal quantity "7" is 00000111.

It is important to remember the difference between a *numerical symbol* and a *number*. In computer languages such as *Visual Basic*, this distinction requires the use of the VAL ("value") statement to convert a symbol into a number. For example, one common operation involves a user entering a number into a text box. The computer reads the text box (which contains symbols for numbers) and converts it into a binary number equivalent. This could be a line of code such as *MyNumber = Text1*. Text fills the variable MyNumber with a text string of the contents of textbox TEXT1. But if you want to convert it to a number, the coding would be MyNumber = VAL(Text1.Text).

On a lower level, if you are programming simple computers or controllers and want to send a "7" to a display, then you must convert 00000111 to 00110111 or confusion will exist (00110111 in standard keyboard language is the BEL — bell ringing — symbol).

Hexadecimal Code

A binary code of four bits can be used to form a hexadecimal (base 16) number. This greatly simplifies entering data or instructions into a digital machine. Entering 7F, for example, is a lot easier than entering all eight binary digits (or 01111111) needed in the binary representation. Using hexadecimal notation permits four bits to be entered at a time. Let's say that we have an eight-bit binary number. The pattern could represent all 256 states from 00000000_2 (0_{10}) to 11111111_2 (255_{10}). Consider this binary number: 10111101_2 (189_{10}). To enter this number on a bit-by-bit basis would mean using eight separate keystrokes: 1-0-1-1-1-1-0-1. But we can break the number 10111101 into two four-bit groups:

1011 1101

Because binary 1011_2 (11_{10}) is hexadecimal number B_{16}, and 1101_2 (13_{10}) is hexadecimal number D_{16}, the hexadecimal number BD_{16} can represent the binary pattern 10111101.

The hexadecimal system can also represent binary words that have bit lengths that are not divisible by four, provided that we assume that the leading missing digits are zeros. For example, assume the output of a ten-bit analog-to-digital converter is 1000101011_2. We can represent this number with three hexadecimal digits: <u>00</u>10 0011 1011 (filled in missing digits underscored). Because 0010_2 is hex 2_{16}, 0011_2 is hex 3_{16}, and 1011_2 is hex 11_{16}, we can represent 001000111011_2 by $23B_{16}$.

Split Octal Code

We can also represent binary data in the split octal system by grouping the binary digits in groups of not more than three each. For example, the binary code 10111101_2 was represented by BD_{16} in the previous section. This same binary number can be divided into octal groups:

010 111 101

Fill in any leading zeros that are needed at the far left digit of the numbers. Because binary 010_2 is also octal 2_8, binary 111_2 is octal 7_8, and binary 101_2 is octal 5_8, the octal number 275_8 produces the same binary bit pattern as the binary number 10111101_2. In this case, however, only three keystrokes are needed (2-7-5) instead of eight (1-0-1-1-1-1-0-1). Both octal and hexadecimal notation can represent the numerical, alphabetic and control codes of systems that are commonly used on keyboards, printers, displays and so forth.

Binary Coded Decimal (BCD)

The *binary coded decimal* (BCD) code is used to represent the ten digits of the decimal number system in a four-bit format. *Table 3-1* shows the standard BCD codes.

BCD is a binary notation that is compatible with computers, yet can represent decimal digits. BCD words are therefore grouped the same way as other forms of decimal digits. The actual value of any BCD digit is dependent on its position with respect to the

Decimal	Binary Coded Decimal
0	0000
1	0001
2	0010
3	0011
4	0100
5	0101
6	0110
7	0111
8	1000
9	1001

***Table 3-1**. Binary coded decimal (BCD) system.*

Decimal	BCD Equivalent
2	0010
23	0010 0011
175	0001 0111 0101

Table 3-2. *Examples of BCD notation.*

other digits. *Table 3-2* gives a couple of examples of how positional weighted notation can be used with BCD.

The BCD numbers are difficult to use in calculations, although it is not impossible to use them. The main use for BCD is in systems requiring a numerical output display, such as a frequency counter, digital panel meter, digital clock, or a scientific instrument. There are a number of BCD converter chips on the market that will convert the four-bit BCD input words to whatever code is needed by the display device.

Excess-Three Code

Decimal	BCD Code	Excess-3 Code
0	0000	0011
1	0001	0100
2	0010	0101
3	0011	0110
4	0100	0111
5	0101	1000
6	0110	1001
7	0111	1010
8	1000	1011
9	1001	1100

Table 3-3. *Excess-3 code.*

The excess-three code is a code formed by adding three to the BCD number. For examples, see *Table 3-3*. Excess-three code is used to make digital subtraction in BCD a little easier. Recall from the previous chapter that *complement arithmetic* is often used in binary arithmetic operations. This cannot be done in straight BCD, however, because it would occasionally result in a disallowed bit pattern that is not recognizable as any of the ten BCD numbers. Excess-three code is nearly obsolete, although it has been seen recently in simple microcontroller-based electronic instrumentation.

Gray Code

The Gray code is used on shaft encoders and other applications that work best if the binary code changes only one bit at a time for each change of state. An example of the Gray code is shown in *Table 3-4*. The Gray code is used most often where a transducer on a mechanical device is used to directly generate data for a computer or another digital circuit (for example, a shaft angle or position in the shaft encoder).

Alphanumeric Codes

Several different codes have been used over the years for representing alphanumeric characters. Obsolete codes include the Baudot code, used on teletypewriters from the 1930s into the 1960s, and the EBCDIC (extended binary coded decimal interchange code) used on IBM mainframe computers in past decades. Today, the ASCII code (American Standard Code for Information Interchange) is used almost universally to represent alphanumeric characters. *Table 3-5* shows the ASCII code.

Decimal	Binary	Gray Code
0	0000	0000
1	0001	0001
2	0010	0011
3	0011	0010
4	0100	0110
5	0101	0111
6	0110	0101
7	0111	0100
8	1000	1100
9	1001	1101
10	1010	1111

***Table 3-4**. Gray code.*

Hex Code	Meaning	Comments
00	NULL	Null
01	SOH	Start of heading
02	STX	Start text
03	ETX	End text
04	EOT	End of transmission
05	ENQ	Enquiry
06	ACK	Acknowledgement
07	BEL	bell
08	BS	Back space
09	HT	Horizontal tab
0A	LF	Line feed
0B	VT	Vertical tab
0C	FF	Form feed
0D	CR	Carriage return
0E	S0	Shift out
0F	SI	Shift in
10	DLE	Data link escape
11	DC1	Direct control 1
12	DC2	Direct control 2
13	DC3	Direct control 3
14	DC4	Direct control 4
15	NAK	Negative acknowledgement
16	SYN	Synchronous idle
17	ETB	End of transmission block
18	CAN	Cancel
19	EM	End of medium
1A	SUB	Substitute
1B	ESC	Escape
1C	FS	Form separator
1D	GS	Group separator
1E	RS	Record separator
1F	US	Unit separator
20	(special)	-------
21	!	-------
22	"	-------
23	#	-------
24	$	-------
25	%	-------
26	&	-------
27		-------
28	(	-------
29	)	-------
2A	*	-------
2B	+	-------
2C	,	-------
2D	-	-------
2E	.	-------
2F	/	-------
30	0	-------
31	1	-------
32	2	-------
33	3	-------
34	4	-------
35	5	-------
36	6	-------
37	7	-------
38	8	-------
39	9	-------
3A	:	-------
3B	;	-------
3C	>	-------
3D	=	-------
3E	<	-------
3F	?	-------
40	@	-------
41	A	-------
42	B	-------
43	C	-------
44	D	-------
45	E	-------
46	F	-------
47	G	-------
48	H	-------
49	I	-------
4A	J	-------
4B	K	-------
4C	L	-------
4D	M	-------
4E	N	-------
4F	0	-------
50	P	-------
51	Q	-------
52	R	-------
53	S	-------
54	T	-------
55	U	-------
56	V	-------
57	W	-------
58	X	-------
59	Y	-------
5A	Z	-------
5B	[	-------
5C	/	-------
5D	]	-------
5E	.	-------
5F	|	-------
60	^	-------

Table 3-5*. ASCII code.*

61	a	-------
62	b	-------
63	c	-------
64	d	-------
65	e	-------
66	f	-------
67	g	-------
68	h	-------
69	i	-------
6A	j	-------
6B	k	-------
6C	l	-------
6D	m	-------
6E	n	-------
6F	o	-------
70	p	-------
71	q	-------
72	r	-------
73	s	-------
74	t	-------
75	u	-------
76	v	-------
77	w	-------
78	x	-------
79	y	-------
7A	z	-------
7B	{	-------
7C	|	-------
7D	}	-------
7E	~	-------
7F	DEL	Delete

Table 3-5. *ASCII code (continued).*

Chapter 4

Digital Logic Families

What is a "logic family?" A digital logic family is any group of digital integrated circuit devices (using similar technology and possessing similar attributes) that work together using direct interconnection rather than elaborate external interfacing, level translation or impedance matching circuitry.

Even in the days when digital logic functions were implemented in discrete components, there were families of circuits. In the integrated circuit, era the concept of "logic family" was formalized.

Early on, there was *resistor transistor logic* (RTL) and *diode transistor logic* (DTL), both of which are now obsolete. In the 1960s and 1970s, the ubiquitous *transistor-transistor logic* (TTL) and *complementary metal oxide semiconductor* (CMOS) emerged. These, and more recent variations on the same themes (such as NMOS, PMOS and Schottky TTL), are the mainstay of digital logic circuits today.

For very high frequency applications, you will also see *emitter coupled logic* (ECL) devices. The front ends of modern gigahertz digital frequency counters are usually ECL devices, even though CMOS or some other form of logic is used further in the circuit.

Why are there so many logic families? Each of the currently popular device types is popular in the market because it has its own place, and electronic circuit designers make decisions based on what they perceive as the most reasonable trade-offs between the vices and virtues of the respective families. It is often said that CMOS devices can do anything a TTL device can do, but at a small fraction of the current drain (microamperes rather than milliamperes). However, this overlooks some facts. For example, there are TTL chips that can provide 80-100 MHz counting, which a CMOS chip can't touch.

Speed and power consumption tend to be directly proportional to each other in digital devices. High speed usually means increased power consumption. It seems that low power operation usually means higher internal resistances and capacitances. This makes the R-C time constants of the circuit longer, thereby accounting for the slower speeds. When you reduce the impedances inside the device (as well as the interelectrode capacitances) to make the device run faster, you also increase the current drain. Hence, you also increase the power consumption.

Some Definitions

Before proceeding to a discussion of the various logic families, let's first take care of a little housekeeping. Some definitions need to be in place first: *current source*, *current sink*, *fan-out* and *fan-in*.

Current Source. A current source is a circuit that will *supply current to a load*. If you connect a resistor to one side of a DC power supply, it then becomes a current source to circuits connected between the free end of the resistor and the other side of the DC power supply.

Current Sink. A current sink is the opposite of a current source. It is a circuit that accepts current from another circuit. In the simplified example above, the load connected to the current source will act as a current sink.

Fan-In/Fan-Out. Most IC logic families use standardized input and output terminals. They are matched according to their roles as current sinks or current sources in the case of TTL. The TTL input, for example, acts as a current source. The TTL output, on the other hand, acts as a current sink. When an output is used to drive an input, therefore, the current produced by the input terminal will find a path to ground through the output of the driving device.

No matter how else you view them, IC digital devices are still little more than a collection of transistors, diodes and resistors. Just as in any direct-coupled transistor circuit, one may pay attention to the drive levels, impedance matching, voltage and current levels at the interface between the two stages. In most analog electronic circuits, the designer must do this chore each and every time a new circuit is designed. But in digital electronics, there is a method for specifying input drive requirements and output drive capabilities. The popular TTL family, for example, has standardized a +5VDC power supply for all devices. An output might produce as much as +5VDC, although the more common range is +2.4 to +3.5 volts. All input terminals must be capable of accepting that level. Similarly, the properties of a standard TTL input are specified in terms of the current source level in milliamperes. The basic unit used in this specification, therefore, is the standard TTL input. With this in mind, the terms *fan-in* and *fan-out* are understandable. The "fan" number specifies the number of TTL standard inputs in question. For example, if the output drive rating is a fan-out of 10, then the device will successfully drive up to ten standard TTL inputs. The fan-in is the inverse of fan-out. It specifies how many TTL input loads this particular input requires. Most devices have a fan-in of one on each input. The designer merely adds up the total fan-in at a point and then makes sure that the driving device has a fan-out that is equal to, or greater than, this number.

TTL Logic

Transistor-Transistor Logic (TTL or T^2L). This was the most popular logic family for several decades. These devices carry type numbers in the 74xx and 74xxx series. Some subfamilies have numbers like 74yxx and 74yxxx, where the "y" is one or two letters denoting the particular subfamily.

Military TTL. There are also 54xx and 54xxx devices out there. These are the military equivalents to the 74xx and 74xxx devices (the "xx" and "xxx" are the same). A 54190 device is merely a 74190 in uniform.

Military digital ICs are typically selected from the ceramic packaged commercial devices, and then tested for a temperature range of -25 degrees C to +125 degrees C. In recent times, however, most semiconductor manufacturers have gotten away from the screening requirements of the military specifications, and refuse to offer such products. At the same time, however, the reliability of plastic chips has increased a tremendous amount. This improvement is due to manufacturing and material changes that were driven by the requirements of the burgeoning electronics technology market. Today, there is considerably less difference between MIL and non-MIL devices than was true twenty or thirty years ago.

TTL devices operate from a +5VDC power supply. In most TTL devices, the DC power supply must be maintained in a small range around +5 volts. The generally given range for all TTL devices is +4.75 volts to +5.2 volts. Performance is unspecified and unpredictable below the lower limit, and damage to the device can be expected above the upper limit. Some people prefer more narrow ranges. For example, the failure rate of TTL devices tends to increase at voltages over +5.05 volts, even though those values are less than the +5.2V upper limit. Similarly, some complex function devices behave erratically below about +4.9 volts, even though the voltage is above the minimum value of +4.75 volts.

The increased operating speed (15 to 25 MHz is typical) is achieved at the expense of increased current consumption. The typical standard TTL device requires 15 to 30 milliamperes. If 22 mA is taken as average, the current requirements rapidly add up. Fewer than fifty integrated circuits will draw one ampere of current from the +5VDC power supply. Many digital circuits have many more than fifty IC devices. Add to that the fact that optimum circuit performance often requires a DC power supply capable of delivering a lot more than "just enough" current, and you will see why high-current 5VDC power supplies were used with TTL-based digital instruments. Fortunately, newer families and microprocessors have reduced the need for "slugger" power supplies.

Typical TTL inverter circuits are similar to *Figure 4-1* and *Figure 4-2*. The device shown in *Figure 4-1* has a typical TTL output. The circuit in *Figure 4-2* has an "open-collector" output stage. The open-collector circuit requires an external pull-up resistor, but this feature allows the device to interface directly with lamps, light-emitting diodes (LEDs), relay coils and logic circuits operating at non-TTL voltage levels.

The regular TTL output circuit shown in *Figure 4-1* uses a totem-pole final stage; transistors Q3 and Q4 form a series-connected pair. Notice that the input of a TTL

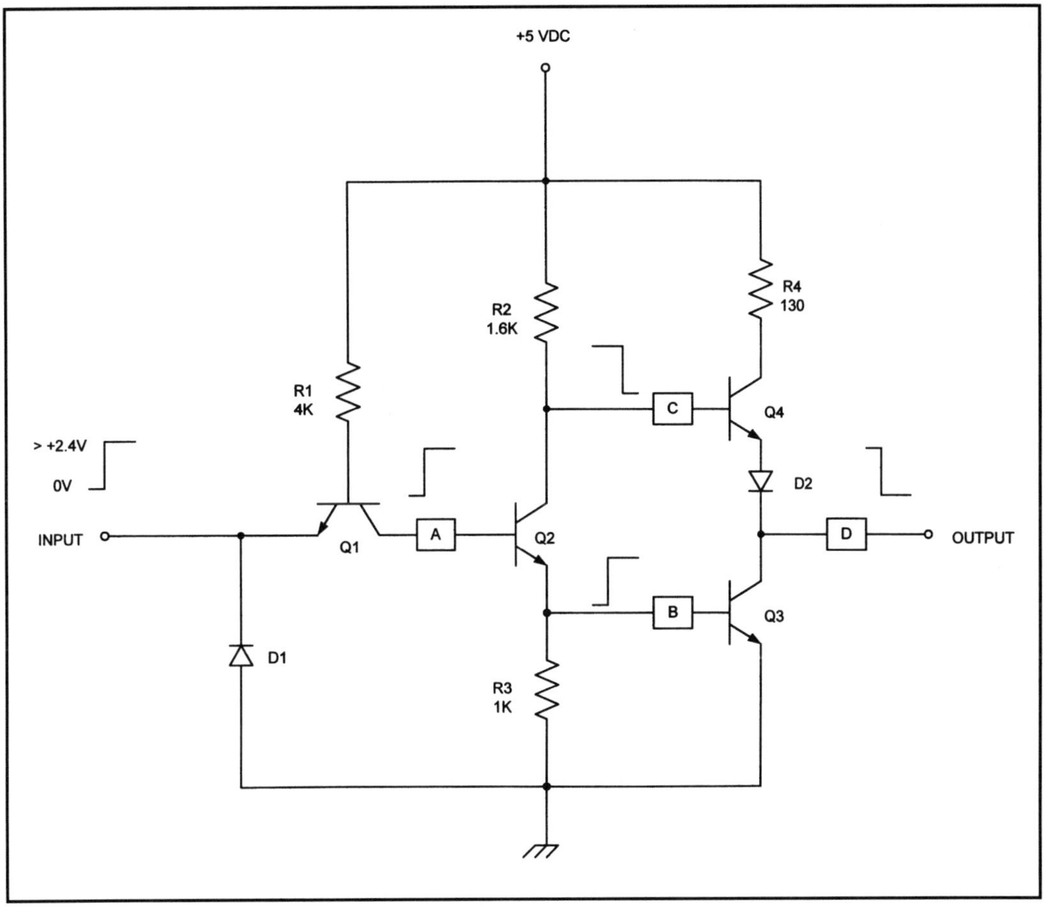

Figure 4-1. TTL inverter circuit.

device is a transistor emitter terminal. The circuit or device (which could be another TTL IC) that is connected to the TTL input must be able to sink the emitter current of Q1 (typically 1.8 mA). It can be said, then, that TTL inputs are current sources, while TTL outputs are current sinks. The operation of *Figure 4-1* follows.

When the input is LOW:

1. The emitter of Q1 is grounded, so Q1 is forward biased.

2. Resistor R1 and the c-b junction of transistor Q2 form the collector load of transistor Q1. Point A, which is where the collector of Q1 and base of Q2 are connected together, is also LOW.

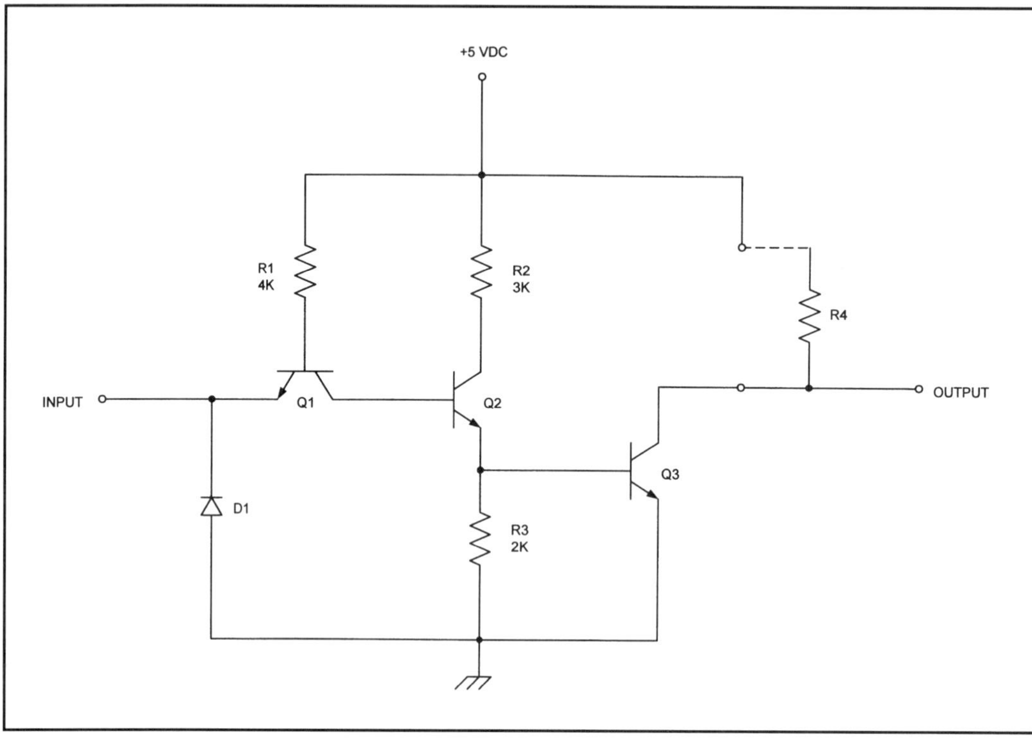

Figure 4-2. *Open-collector TTL inverter circuit.*

3. When point A is LOW, Q2 is cut off, making point B LOW and point C HIGH.

4. Under this condition, Q3 is turned off and Q4 is turned on, connecting the +5VDC supply to the output terminal.

When the input is HIGH:

1. Transistor Q1 is turned off, so point A is HIGH.

2. When point A is HIGH, transistor Q2 is turned on, making point B HIGH and point C LOW.

3. When point C is LOW, Q4 is turned off, disconnecting the +5V supply from the output. Point B is HIGH, so Q3 is turned on. This causes the output terminal to be at ground potential.

An inverter circuit such as *Figure 4-1* needs no external components to operate. But that feature also makes it more difficult to interface the device with anything other than transistors or other TTL IC devices. In many cases, therefore, an open-collector TTL device is required. The open-collector TTL inverter is shown in *Figure 4-2*. Notice that the input and driver stages (Q1/Q2) are essentially the same as in the previous case. But only Q3 is used in the output stage, and it has no collector circuit. The collector of Q3 is brought to the output world through a terminal and must be returned to a positive DC voltage through an external load (such as resistor R4). The supply end of R4 must go to +5VDC in some cases, while it may go to any voltage between either +5VDC and either +15VDC or +30VDC, depending on the device type number. An open collector TTL IC would be used to turn on a lamp, LED, relay and so forth.

The multiple input gates of the TTL series are similar to the circuits shown previously, but use a multiple emitter input stage (*Figure 4-3*). Each TTL input must be a transistor emitter terminal.

Emitter Coupled Logic (ECL)

Emitter-coupled logic (ECL) is a high-speed IC logic family of nonsaturated devices. "Nonsaturated" means that the HIGH and LOW levels are not hard against the DC power supply rails. TTL, on the other hand, operates with the transistors in saturation or cut off all of the time. Operating frequencies to 120-MHz are common,

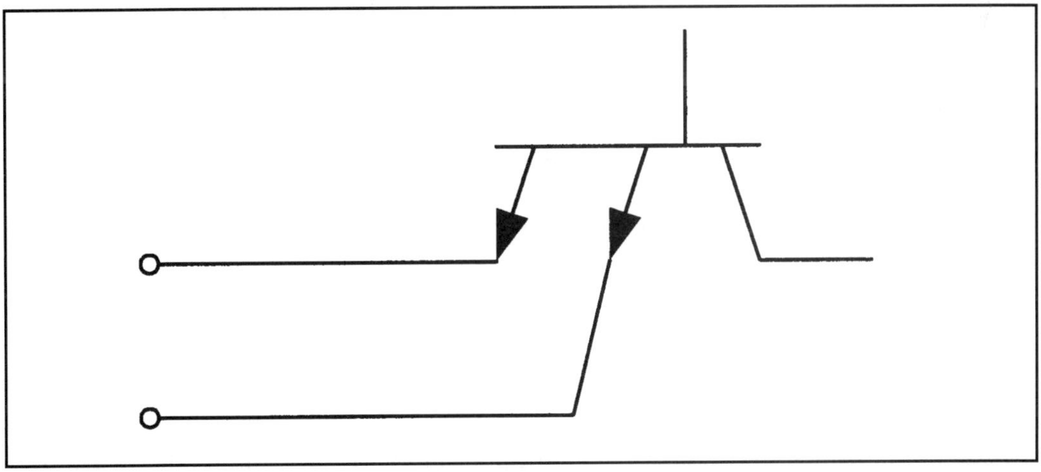

Figure 4-3*. Multiple input TTL device uses additional emitters.*

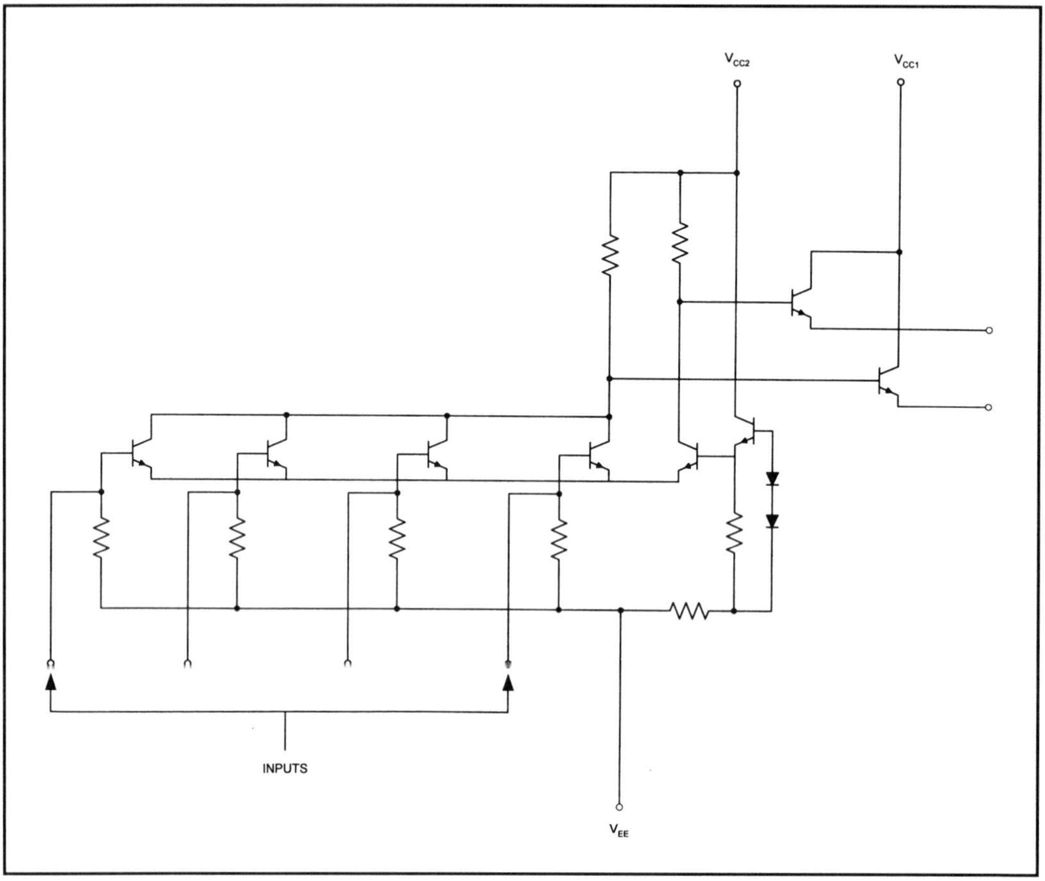

***Figure 4-4**. ECL device.*

and 2-GHz frequencies are possible. The higher end devices are often used as prescalers for digital frequency counters. The actual counter may be a 100-MHz counter, but with a 1-GHz ECL prescaler, operation is expanded to 1,000-MHz.

Figure 4-4 shows a TTL gate circuit. This particular device is a four-input OR/NOR gate. Note that both inverted (NOR) and noninverted (OR) output terminals are available.

These devices offer optimum AC performance when VCC and VEE are set to ± 5VDC, respectively. The ECL family combines very high speed with moderate power consumption, something usually deemed impossible for the saturated logic families (TTL and CMOS).

High Threshold Logic

The TTL logic devices that have become standard in the electronics industry have only 400 to 500 millivolts of noise immunity, or the ability to not respond to noise signals. In many applications, this is not sufficient to eliminate false operation. An industrial site that uses lots of electrically operated machinery is considered electrically noisy. Pulses picked up from the electrical activity will be treated as valid pulses by TTL logic elements.

The *high threshold logic* (HTL), also called *high noise immunity logic* (HNIL), offers superior immunity to spurious noise pulses. The first HTL devices were designed to operate from +12VDC power supplies, while later devices operated at +15VDC power supplies. CMOS devices are technically "high threshold" devices, but are not included in this discussion because they are sufficiently different (and more common) to warrant separate discussion. Bipolar HTL devices have approximately ten times the noise immunity of TTL. The improved noise immunity is derived from three factors:

1. Lower circuit impedances.
2. Slower operating speeds.
3. High threshold operating potentials.

Additionally, HTL devices usually offer a ±1V power supply immunity, as opposed to the ±250 mV tolerance of TTL. As a result, some circuits using HTL devices do not require regulation of the DC power supply.

Figure 4-5 shows the relative noise immunity of the two principal HTL families (+12V and +15V), along with the standard +5V TTL family. Notice that TTL devices have a 400 mV (0.4V) noise immunity band. The 12V HTL family has a 3.5V noise immunity band, almost ten times better. The 15V family, however, does even better, having a 3.2V band at the lower end and a 6.5V band at the upper end.

Figure 4-6 shows a typical HTL gate. Notice that it is in some respects similar to the obsolete DTL circuits, except for the 5.8V zener diodes in series with the input diodes. The zener diode operates to raise the voltage threshold required to operate the gate. An input signal must be high enough in level to exceed the zener breakdown potential of the diodes, or the gate will not respond.

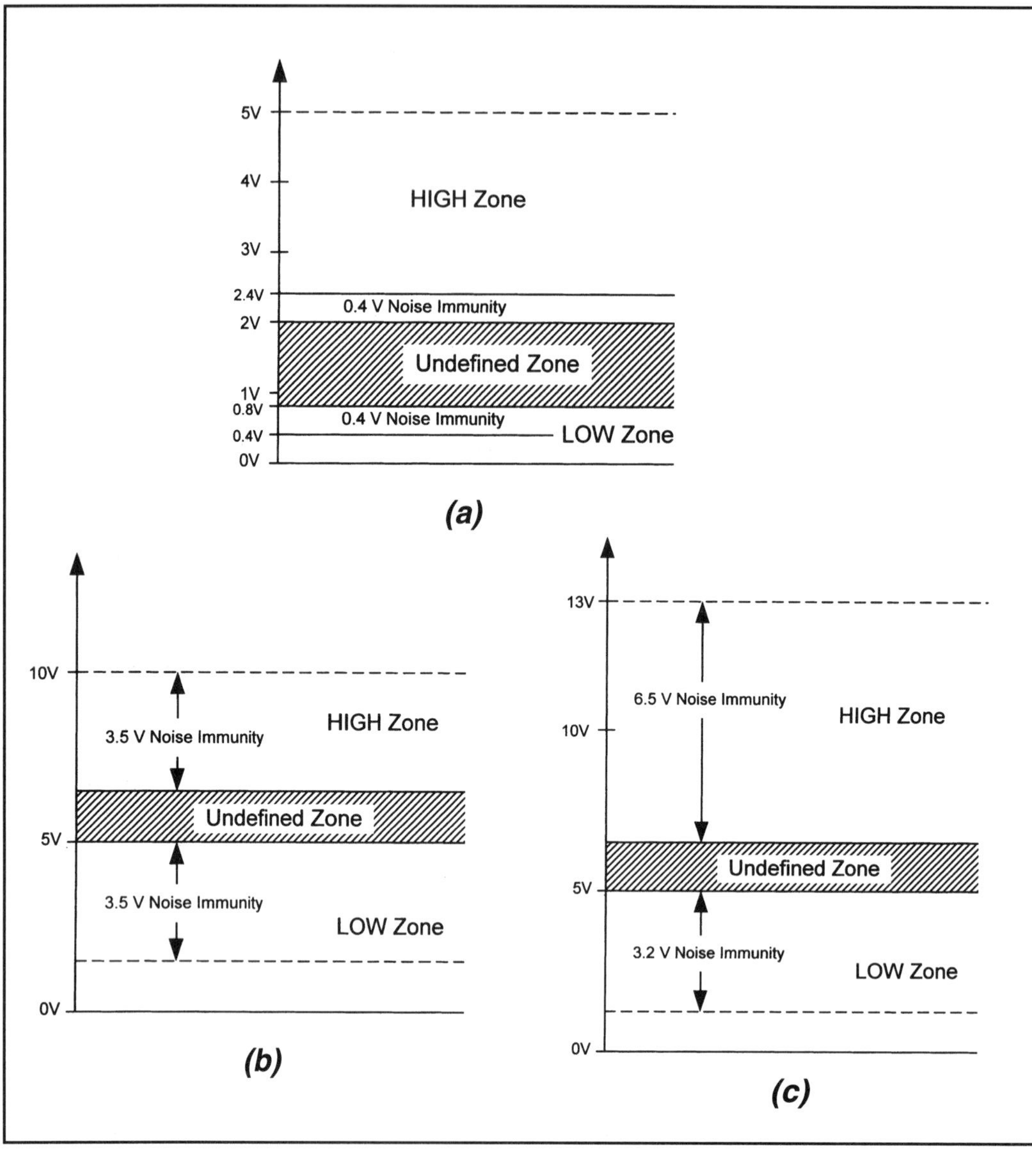

***Figure 4-5**. a. TTL noise immunity; b. 12V HTL noise immunity; c. 15V HTL noise immunity.*

HTL inputs are current sources, so any device driving an HTL input must be able to sink a current equal to approximately 2.1 mA in 12V devices, and 2.6 mA in 15V devices.

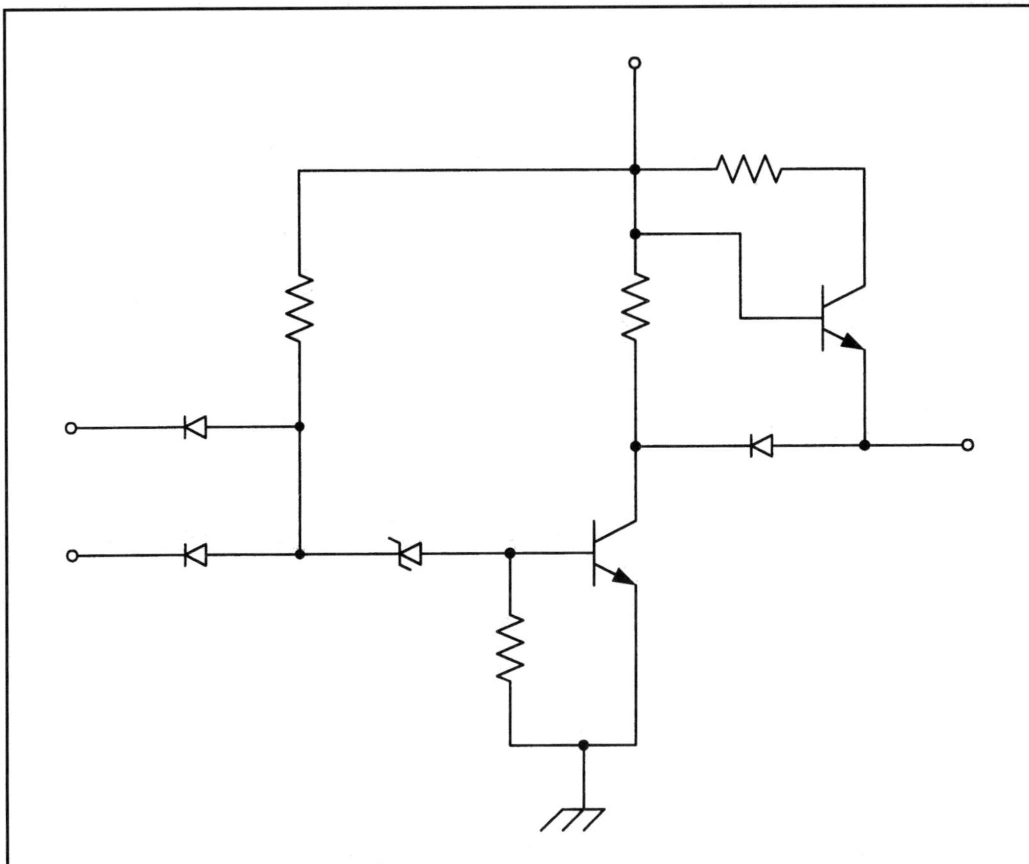

Figure 4-6. HTL inverter device.

HTL devices can be interfaced with elements of other digital logic families. They can be interfaced directly with TTL device inputs, provided that the +VDC power supply of the HTL device is the same potential as used in TTL devices (such as +5VDC). Interfacing with HTL devices operated at other potentials, however, may require some external components. To interface an HTL output with a TTL input, an HTL device that has an open-collector output is usually needed. A 3 kohm to 4 kohm pull-up resistor can then be connected between the open collector output terminal on the HTL device to the +5VDC power supply of the TTL device.

Similarly, TTL open collector outputs can be interfaced with HTL inputs if a 10 kohm pull-up resistor to the +12VDC or +15VDC power supply is provided. A restriction on this approach, however, is that the TTL open collector must be a +15V or +30V type, such as the 7406 inverter.

CMOS Devices

Complementary metal oxide semiconductor (CMOS) logic elements are newer than TTL, but they are now close to dominating the market. It's a little premature to write the obituary of TTL and ECL devices, however. CMOS devices differ from the other families in that they use MOSFET field effect transistors instead of PNP and NPN bipolar transistors. It is a saturated logic family that is well regarded for its low current consumption.

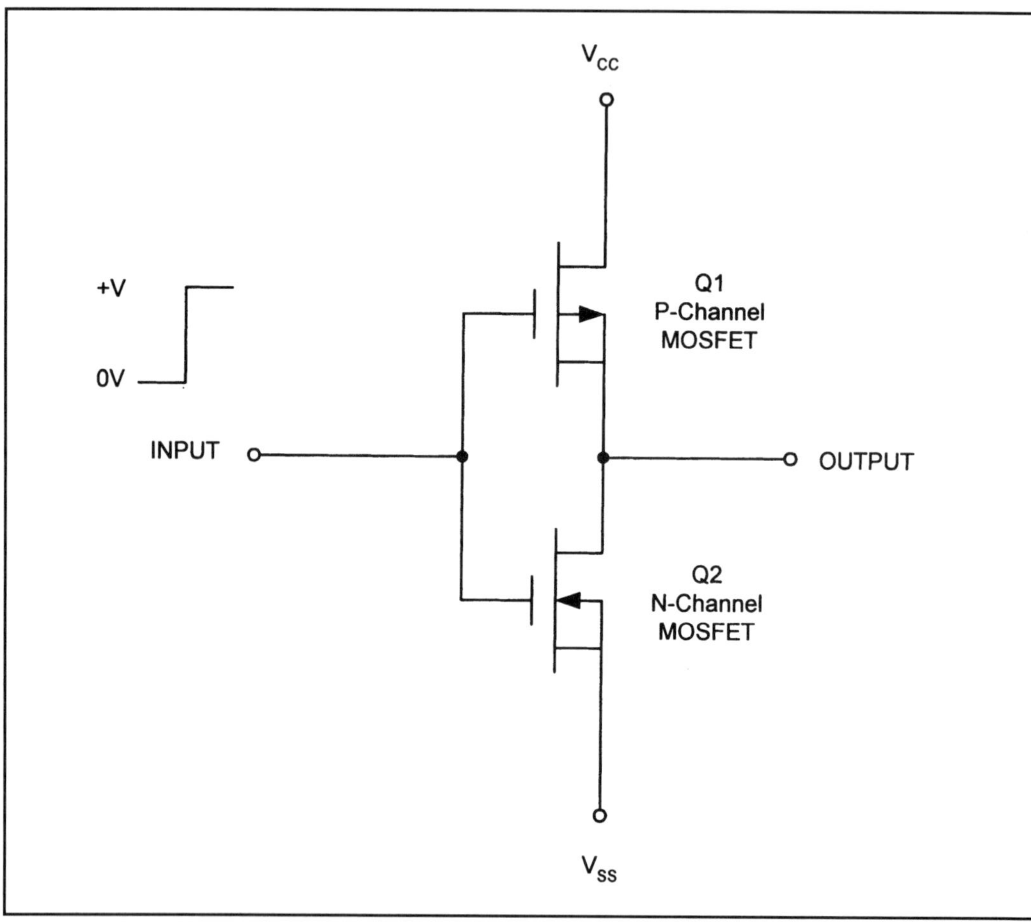

Figure 4-7. *CMOS inverter.*

It takes two types of MOSFETs to make a CMOS inverter stage (*Figure 4-7*): a p-channel MOSFET and an n-channel MOSFET. The p-channel device turns on when its gate potential is zero or negative with respect to the source. The n-channel turns on when the gate is positive with respect to the source. That's the complementary part of it.

Figure 4-7 shows a basic CMOS inverter circuit that is designed with n-channel and p-channel MOSFETs connected in series. The output is taken from their mutual junction, the source end of Q1 and the drain end of Q2. The gates are connected in parallel, so the same potential is applied to both Q1 and Q2.

The series configuration means that the channel resistances are in series. Because of the complimentary nature of the circuit, one resistance will be very high when the other is very low. The potential appearing at the output terminal is the result of the voltage divider action of the two resistances on VCC and VSS.

When the input signal is LOW (0 volts or some negative potential — both can be used in CMOS devices), transistor Q2 is turned off and Q1 is turned on. This allows the V_{cc} potential to appear at the output terminal. In this condition, only a very small resistor from Q1 is in series with the very high resistance of Q2.

When the input signal goes HIGH, on the other hand, exactly the opposite situation occurs: Q1 is turned off and Q2 is turned on. In this condition, there is a high resistance between the output terminal and V_{cc}, and a low resistance between the output terminal and V_{ss}.

A model of this action is shown in *Figure 4-8*, using resistors to model the action of the channel resistances. The circuits are resistive voltage dividers. In *Figure 4-8a*, the input is LOW, in which case Q2 is off and Q1 is on. This makes R_{Q1} high and R_{Q2} low. Because of the voltage divider action between R_{Q1} and R_{Q2}, the output will be HIGH.

The reverse situation is shown in *Figure 4-8b*. The input is HIGH. In this case, Q1 is off and Q2 is on. This will reverse the resistance relationship between the two channel resistances, so the output is LOW.

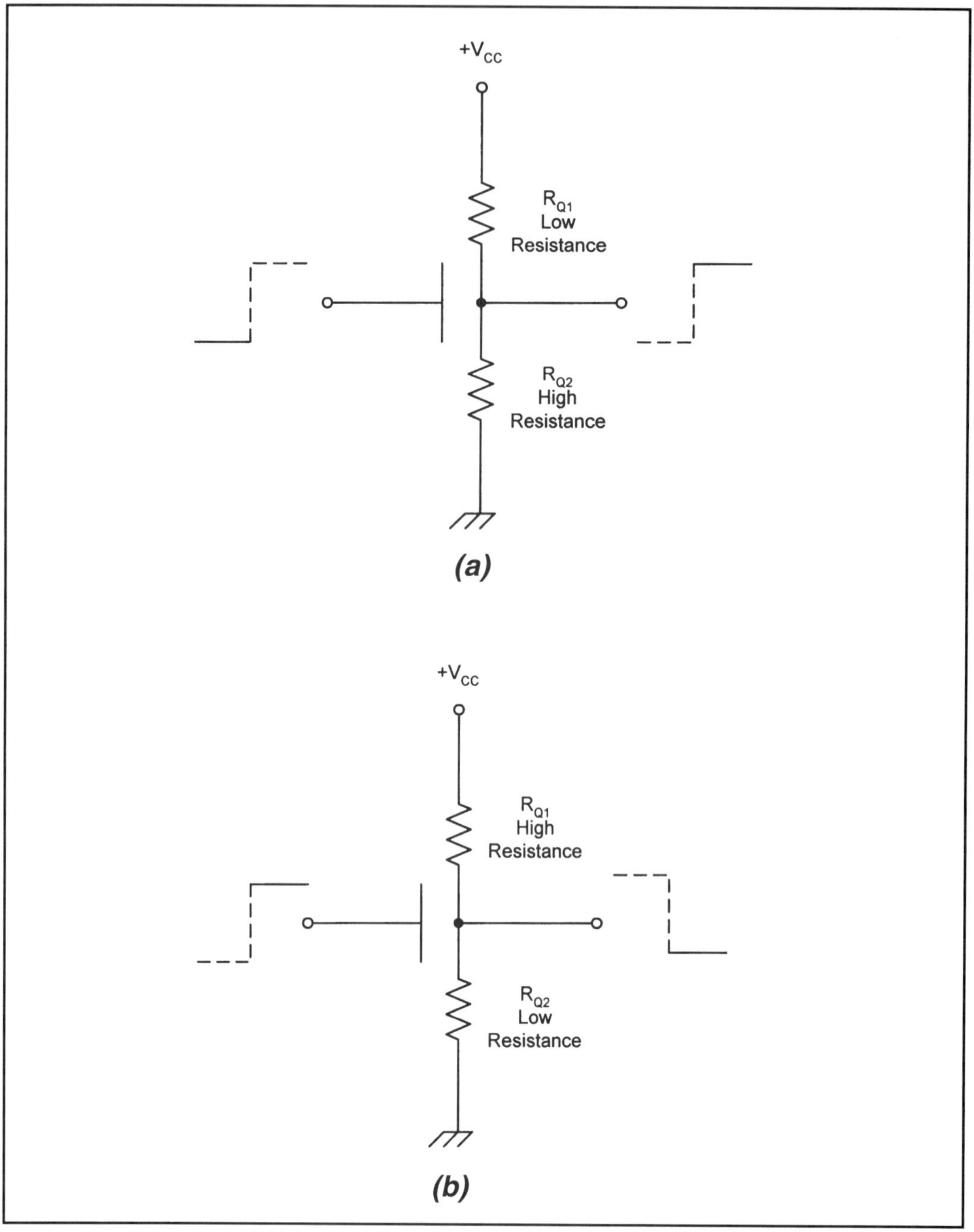

Figure 4-8. *Resistor model of inverter: a. Input LOW; b. Input HIGH.*

The models shown in *Figure 4-8* demonstrate the really unique nature of CMOS devices. These devices will have a low resistance path to VCC in the HIGH input state, and a similar path to ground in the LOW input state. The off transistor in any given state will have a high channel resistance. This situation — a high and low resistance in series — causes the DC power supply to always see a high impedance to ground. The only appreciable current drain occurs when the output is in the middle of its transition state HIGH-to-LOW or LOW-to-HIGH. At those times, the respective on-off roles of the MOSFET transistors are changing to the opposite state. At some time during this transition, the channel resistances will be considerably lower than maximum and will be equal. The current at this instant is maximum, although it still does not approach TTL current levels. During the stable periods, the current is in the microampere range.

How do the two most popular IC logic families compare with each other? Both CMOS and TTL devices have their own adherents who will swear by them (and swear at the other). In truth, though, it must be said that both types of devices have their virtues and drawbacks, and in some areas the design selection can be done by a coin toss. Little or no difference will be noted. In other cases, however, there are clear reasons to use one rather than the other.

When using the TTL device, be ready to accept the fact that TTL ICs draw a considerable amount of current from the power supply. Additionally, there is the fact that the TTL device must operate from a regulated +5VDC power supply that is negative grounded. No other supply will do for TTL devices. CMOS devices, on the other hand, will draw very small amounts of current (microamperes vs. milliamperes), and are able to operate with a wide variety of DC power supply voltages. In many cases, the DC power supply for CMOS devices need not be regulated. Some designers take advantage of the fact that the CMOS device will operate from both positive and negative supplies.

The difference in current requirements between CMOS and TTL is, perhaps, best illustrated by two similar gates. A certain TTL NAND gate requires 8 mA when the gate output is HIGH, and 22 mA when the output is LOW. A CMOS device that is functionally identical, such as a NAND gate, requires only 15 μA in the HIGH condition, and 160 μA in the transition state. It is this low power current drain feature that makes CMOS very popular with designers.

The operating voltage range of the CMOS devices is very wide, while the TTL devices are limited to using a narrowly regulated +5VDC supply. Most CMOS devices will operate properly over the range ±4.5 volts to ±18 volts. In many cases, ±15 volts or ±12 volts are used. The negative power supply is used instead of the needed ground connection when monopolar DC supplies are used. However, most CMOS devices are designed for optimum performance in the 7V-to-12V range. Some complex, special-function CMOS ICs will not operate properly at less than 7 volts, despite being rated down to 4.5 volts.

In contrast to the wide operating range of the CMOS device, we find that the TTL device will not operate well if the power supply voltage deviates very far from the nominal +5V value specified. In fact, many devices are listed as requiring a regulated DC power supply voltage between +4.7 volts and +5.2 volts. It is the experience of many people, however, that many circuits will not operate properly (especially at the higher end of the frequency range) at DC power supply potentials below +4.9 volts. This is especially true of the more complex special function devices, or where there are a lot of ICs in the circuits. Monostable multivibrators and clock circuits especially tend to become troublesome outside the range +4.9 volts to +5.2 volts.

The high current drain requirements of TTL devices often cause problems in power supply distribution networks on printed circuit boards. The copper tracks used to carry the +5 volt and ground return currents will cause I x R voltage drops sufficient to reduce the voltage at a chip below the acceptable operating range. In addition to this current starvation, there is also the problem of *glitches*, or spurious pulses in the circuit. These pulses raise the dickens with gates, counters, flip-flops and monostable multivibrators. A glitch usually consists of a sharp, negative-going pulse on the +5VDC power supply line. It is caused when a sudden current demand occasioned by turning on a chip reduces the terminal voltage at the chip. Any PC board that contains even a modest number of TTL devices requires bypass capacitors (0.001 μF to 1 μF, with 0.1 μF being most commonly used) generously sprinkled all over the board. There should be at least one capacitor for each TTL package, and it should be physically located as close to the +5VDC terminal of the chip as possible. The modern small 0.1 μF capacitors with 0.100-inch lead spacing work best for most applications, although disk ceramic and other forms of capacitor can also be used.

The glitch problem also forces us to use PC boards whenever possible. Regular wiring, wire wrapping, or the plug-in pin socket "breadboarding" forms of construction often create problems due to lead inductance (and in the case of breadboarding sockets, there may also be a serious spurious capacitance present). The inductance of a short piece of wire may not seem like much, but it becomes significant when the rise times of the signal are very fast (as is often the case in TTL circuits) and the current at peak flow is high. Wire connections should only be used with extreme caution in digital circuits.

I recall a fellow in engineering school who used about twenty-five TTL devices in his senior project. The circuit worked somewhat and sometimes, but at other times it went flaky on him. The professor asked me to look at the project for the kid, and the problem immediately was apparent to me: he built it "rat's nest" style with thin wire and no decoupling or bypass capacitors. I had him use a wire-wrap board (second best, but better than perf-board) and place a 0.1 μF capacitor at each chip package. I also had him place a 100 μF at the DC power supply lines, at the end of the DC distribution system, and at the midpoint in the DC distribution system. After that, the circuit worked nicely ... most of the time (there were still design flaws hidden by the construction flaws).

We also find that, in many cases, the noise immunity of the CMOS devices is superior to that of TTL devices of similar function. The TTL device has a few hundred millivolts of noise immunity range, while the CMOS device has several volts. The reason for this difference is that the CMOS device will not change state unless the input signal passes through the VCC/2 point. If a +12V supply is used, for example, the trip point differentiating between HIGH and LOW conditions is 12V/2 = 6V. The TTL device, on the other hand, is not firmly in the LOW state unless the signal level is less than 0.8 volts (800 mV), and is not in the HIGH state unless a signal level is >2.4 volts.

CMOS is a slow logic family compared with TTL. Almost any garden-variety TTL device will operate to 18 MHz, and most will operate to 20 MHz or 25 MHz. Some selected devices or special type numbers operate to 80 MHz. On the other hand, CMOS operates only to 10 MHz or so, and some older devices are limited to 5 MHz.

Another problem with CMOS devices is that they will be damaged by static electricity potentials that inevitably build up on your body, clothing, tools, instruments and the workbench itself. While most CMOS devices are safe to handle when connected into their circuits, this does not, incidentally, give you license to handle the CMOS circuit card carelessly. Damage can and does occur in-circuit. You will hear some technicians and engineers say that the problem is overstated, but it really happens. After having blown out a few terribly expensive devices, and causing damage to low-cost devices that were difficult to locate, I disagree. Always use proper handling procedures.

The CMOS device is sensitive for the same reason that MOSFET transistors are sensitive: low gate-channel breakdown voltage, which is typically around 80 volts. The gate element of a MOSFET device is insulated from the channel by a layer of oxide insulating material that may be as thin as 0.0001 inch. The voltage level of the static electricity on your body is often several thousand volts. This level accounts for the spark and slight shock received when you discharge the static charge into a grounded object.

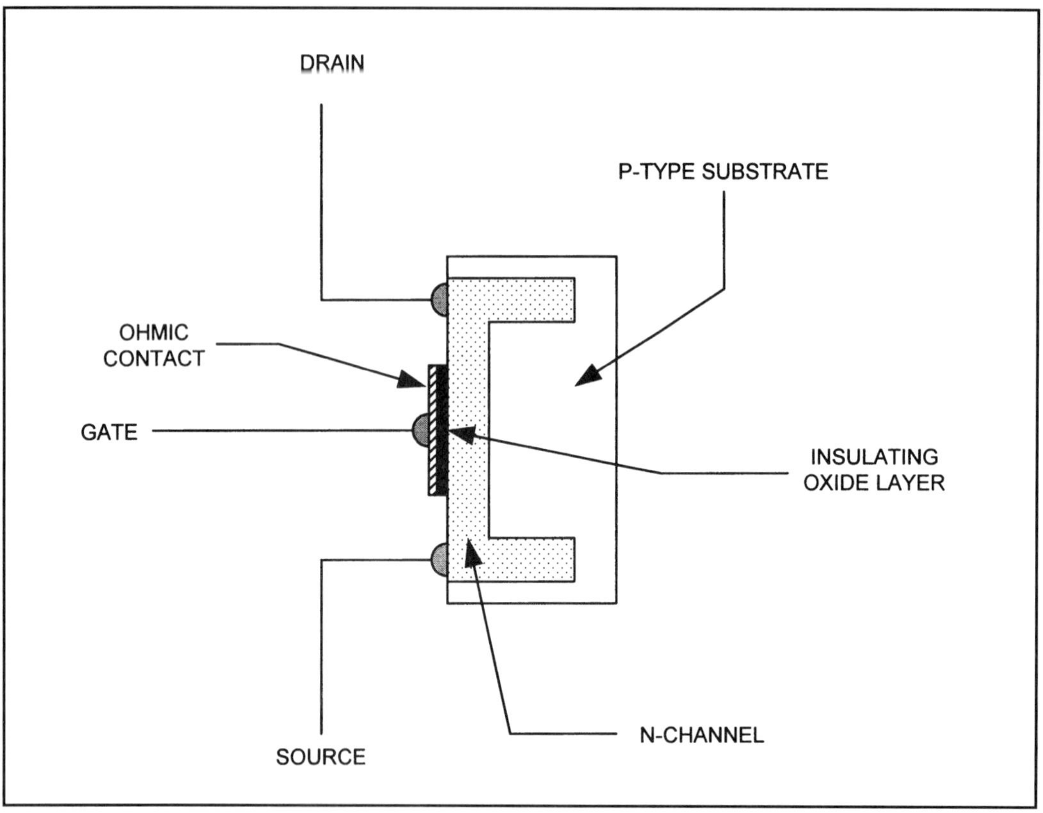

***Figure 4-9**. MOSFET structure.*

Figure 4-9 and *Figure 4-10* show the actual mechanism of CMOS and MOSFET failure. When the integrity of the thin insulating layer is breached, the metal ions in the ohmic contact of the gate terminal are drawn into the hole, creating a low resistance path between the channel and the gate. Unfortunately, the problem may not manifest itself immediately. The breach may not be large enough at first to cause an immediate short circuit; but after several months of operation, there will be enough metallic ions in the breach to cause a short circuit. This may account for the disgusting habit some devices have of failing after a few weeks of operation following a repair action. The cure for this problem is to use special handling procedures and a special working environment in which everything is maintained at the same potential, usually ground potential (0 volts).

Some CMOS devices, such as the inverter circuit of *Figure 4-11*, have internal zener diodes to prevent breakdown. These diodes will clamp the inter-element voltages to some value less than breakdown potential. These diodes are not perfect, however, and cannot be trusted in the absolute sense. The B-series CMOS devices have this type of protection.

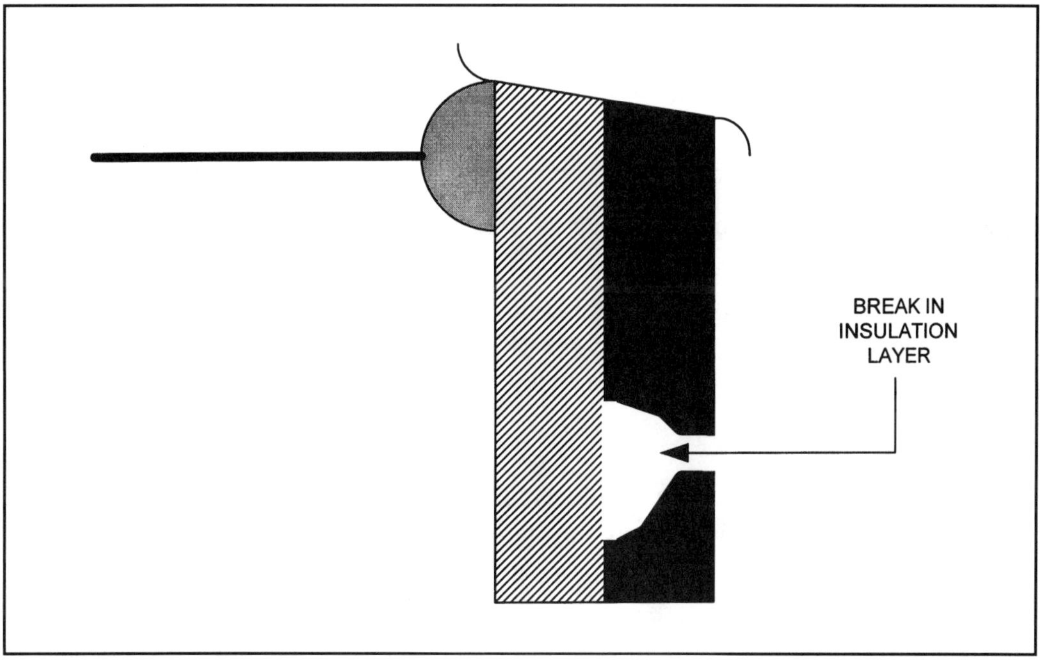

Figure 4-10. *Genesis of ESD damage: broken insulation layer.*

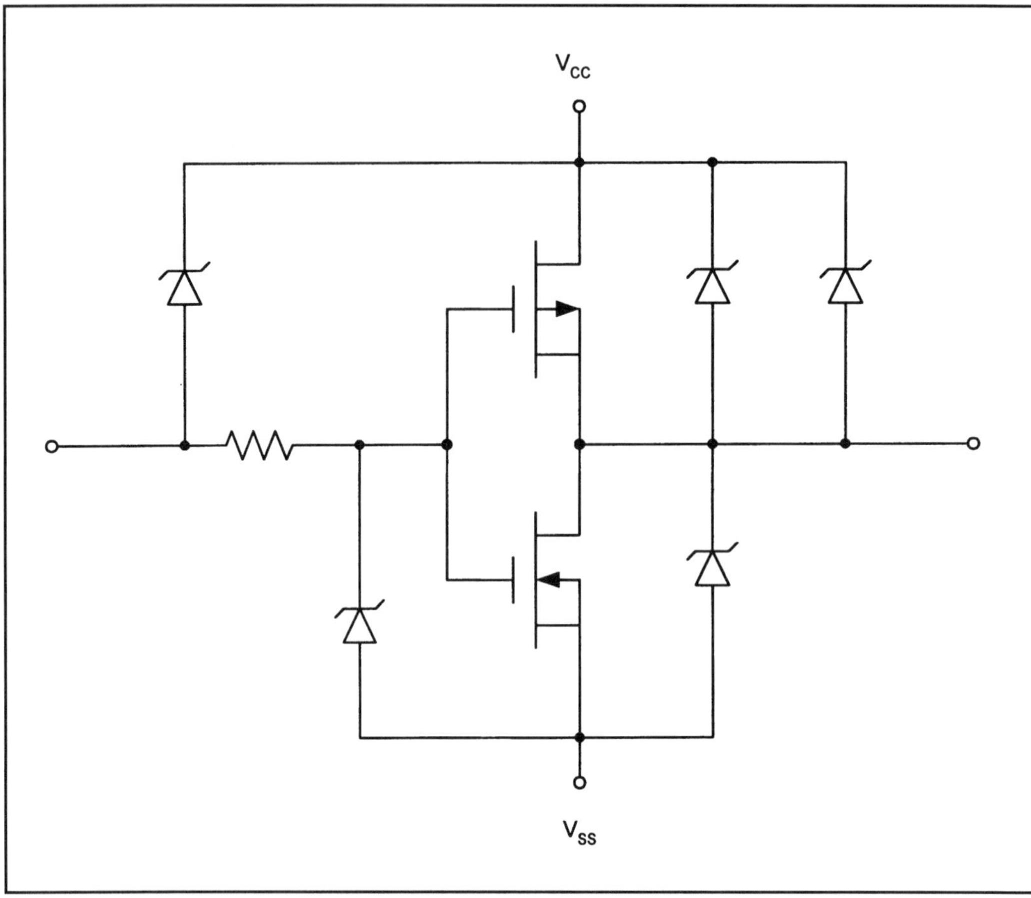

Figure 4-11. *ESD protected CMOS inverter.*

CMOS devices are shipped from the manufacturer and distributor in antistatic containers. They are either conductive plastic or conductive foam, or both. Do not remove CMOS devices from their shipping container or the foam backing until you are ready to install them.

The work environment that you should establish when working on CMOS circuits is a grounded system. This will allow the static electricity to drain to ground before it can do any damage. One tactic is to use a larger layer of sheet aluminum or other conductive material on the work surface. This surface must be kept at ground potential. Similarly, your body and all of your tools and instruments must be grounded. The ground to your body can be an alligator clip load connected to your watch band, but that can be very, very dangerous around 120VAC power supplies. The best form is the high resistance wrist straps and ground leads made especially for the purpose.

Electrostatic Discharge (ESD) Handling Rules

Only a few rules are needed for safe handling of ESD sensitive devices such as CMOS chips. Failure to follow these rules may result in premature, perhaps very sudden, failure:

1. Always handle the device, or printed circuit board containing the device, in an environment grounded to earth through a high resistance.

2. Keep devices and PC boards in their antistatic shipping containers until ready for use.

3. When removing the CMOS device of the PC board from its shipping container, ground yourself through a high-resistance device such as a wrist strap.

4. Before actually touching the device or PC board, touch the chassis of the equipment being prepared to receive the item. This creates an equipotential situation.

5. If you lay the device or PC board on the bench, lay it on a sheet of conductive material.

6. Always avoid touching the IC pins any more than is necessary.

7. Use only soldering irons that have grounded tips (look for a three-wire AC line cord).

8. Turn off the power before inserting the device into the circuit. (Do this anyway, regardless of ESD sensitivity.)

TTL and CMOS Subfamilies

There are a number of subfamilies or related families of digital IC logic devices. There is also a number of CMOS family devices that have TTL outputs, so carry a "74xxx" type number.

Type Numbering Scheme

The general type numbering scheme is NNYYyyZzzzz, where:

XX = 74 for commercial TTL or TTL-output CMOS devices.
XX = 54 for military grade devices.
XX = 75 for interface devices.

Yyy = {null} or N for traditional TTL.
Yyy = L for low-power TTL.
Yyy = H for high-power TTL.
Yyy = S for Schottky TTL.
Yyy = LS for low-power Schottky TTL.
Yyy = AS for advanced Schottky TTL.
Yyy = ALS for advanced low-power Schottky TTL.
Yyy = F for fast TTL.
Yyy = C for CMOS with TTL output.
Yyy = HCT for high-speed CMOS with TTL output.
Yyy = HC for high-speed CMOS with CMOS output.
Yyy = AC for advanced high-speed CMOS.
Yyy = ACT for advanced high-speed CMOS with TTL output.

Zzzzz is a four-digit number designating the specific device.

Chapter 5

Gates: The Basic Building Blocks

Several elementary forms of digital logic gates form the basic building blocks out of which even the most complex digital circuits are formed. Even the most complex microprocessor chip, with clock speeds that sound like radar frequencies, is little more than a specific large-scale configuration of the same basic gate elements. There are only a few basic types of logic gate, but they can be combined to form all manner of high-level integrated circuit digital devices. The basic forms of gate are:

NOT gate
OR gate
AND gate
NOR gate
NAND gate
XOR gate

Examples of these gates can be found as stand-alone functions in all of the standard integrated circuit logic families and subfamilies.

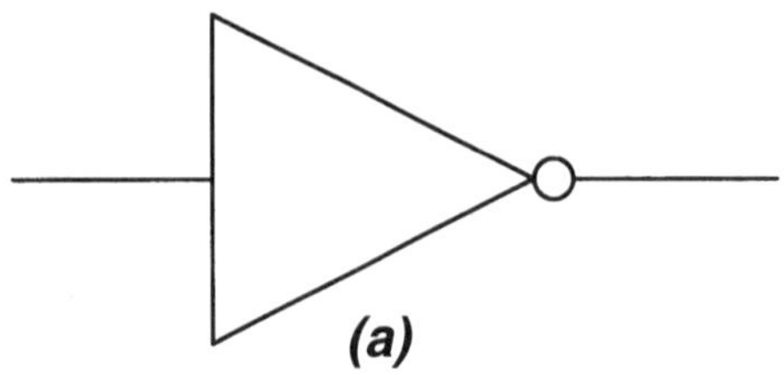

A	$\overline{A}$
1	0
0	1

(b)

***Figure 5-1**. a. Inverter symbol; b. truth table.*

NOT Gates (Inverters)

Figure 5-1 shows the schematic circuit symbol and truth table for the most elementary type of gate. This gate is called by various names: NOT gate, inverter, and complementor.

The action of the NOT gate is simple: the output is always the opposite state as the input. Thus, when the input is LOW, the output is HIGH. And when the input is HIGH, the output is LOW. In other words, the NOT gate inverts the input signal, hence its alternate name, "inverter." The other alternate name, "complementor," comes from the fact that the NOT gate complements the input. In digital arithmetic, the complement of a digit is its opposite. Thus, 1 is the complement of 0, and 0 is the complement of 1. Or put another way, HIGH and LOW are complements of each other.

The term *NOT* comes from the fact that the states are binary in nature, so they can only be either of two possible states: HIGH and LOW. If a signal is HIGH, then it cannot be LOW, or put another way HIGH = NOT-LOW. Or alternatively, LOW = NOT-HIGH.

In some cases, a special notation is used to designate the complemented state. If "A" represents a signal, then an "A" with a bar over it ("bar-A") is the NOT-A. Because of the difficulty in typing such a character, I will retain the denotation "NOT-A" in text, but will occasionally use bar-A in illustrations.

Truth tables are used in discussing digital circuits. Such tables show all possible states and their relationship to one another. For the NOT gate, the truth table is simple:

A	NOT-A
0	1
1	0

This table tells us that whenever input "A" is at logical level 1 (or HIGH), then output "NOT-A" is at logic level 0 (LOW). In the dynamic situation, "A" drops from HIGH-to-LOW, then NOT-A snaps LOW-to-HIGH.

The schematic circuit symbol for the inverter is shown in *Figure 5-1*. It is a triangle with a circle on the output terminal (the input terminal is at the middle of the side opposite the output). Anytime you see a digital logic element symbol with a circle at the output, then it means that the output will have the opposite state than the input, or some other terminal. In the case of the inverter, the output is opposite the input so the circle is at the input. In certain other logical elements an output designated Q will often have a complementary output NOT-Q.

Multi-Input Symbols. Sometimes you will see a multi-input circuit symbol in which the circle is on an input. This means that the inversion occurs only for that input, and not for others.

Buffers. There is another device used in digital circuits called a buffer. It looks like an inverter, but without the output circle that denotes inversion (*Figure 5-1*). These are sometimes also called *noninverting buffers*. Such devices produce an output that is identical to the input (A = A). There are two reasons for using buffers. One is to provide isolation between the circuit from which the signal is received and its destination. The most important use, however, is to increase the drive level of the circuit. The drive capability of a digital circuit is described by its *fan-out*, or the number of inputs of the same type of logic family that the output can drive. A fan-out of 10, for example, means that the output can drive ten standard input terminals. Buffers typically have higher fan-outs than straight gates.

Another use for a buffer is to drive long transmission lines. The digital signals look like pulses to transmission lines, and will deteriorate by the time the signal traverses the line. Using a buffer with a sufficiently high capacity to drive the line will limit this effect.

The device shown in *Figure 5-2* is called a *hex inverter*. It contains six inverters that are totally independent of each other except for sharing common DC power supply

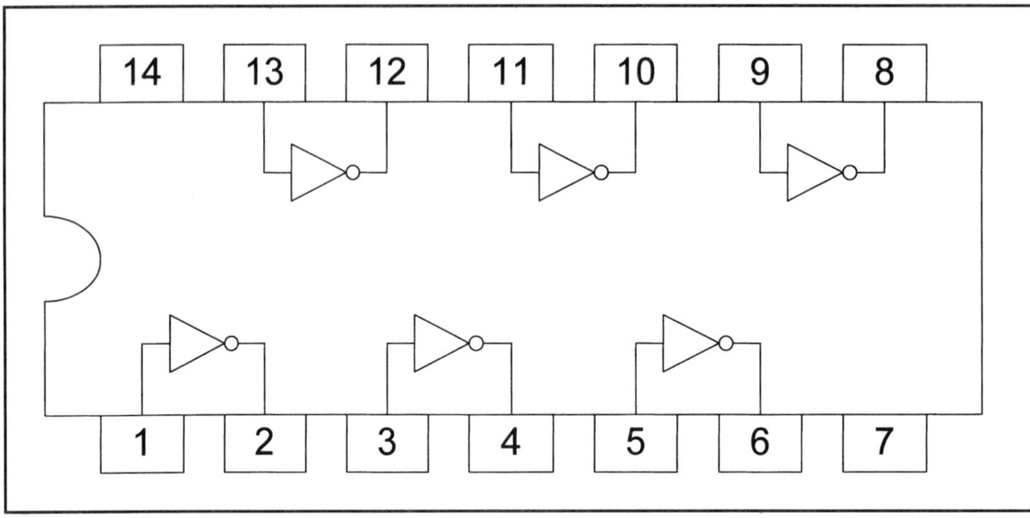

Figure 5-2. *Hex inverter.*

terminals (+5VDC and ground). There is no interaction between the inverters unless external connections are made between the inputs and outputs. On the TTL version of this chip, the average current drain is 12 to 15 mA.

Three related TTL devices use these same pinouts: 7405, 7406 and 7416. All three of these devices are open collector inverters, which means that the output transistor in each inverter requires an external pull-up resistor to the +5VDC power supply. The resistor connects the collector of the output transistor (connected internally to the IC output terminal) to its DC power source. The 7405 is very similar to the 7404 except for the required 2.2 kohm pull-up resistor to the +5VDC power supply.

The 7406 and 7416 devices are open collector devices, but are able to operate at DC potentials up to either 15VDC (7416) or 30VDC (7406). Note that the package voltage applied to pin number 14 must remain at +5VDC. Only the output transistor collector terminal can operate at the higher 15 or 30VDC potentials. The higher collector voltage rating is intended to make it possible for the IC to drive an external load such as a relay coil, lamp or LED.

The 7406 device can sink up to 30 mA of DC current, while the 7416 device can sink up to 40 mA through the output transistor collector. This specification means that the 7406 must have a DC resistance load of not less than R = (30V)/(0.03A) = 1,000 ohms. The 7416 load must be not less than 375 ohms.

Figure 5-3 shows the pinouts for the 4049 and 4069 CMOS hex inverter devices. The 4049 device may be used as a level translator between +5VDC TTL devices and those CMOS devices that are operated at potentials other than +5VDC. The voltage applied to pin number 1 defines the output voltage swing; so if +5VDC is used, the device becomes TTL compatible, even though it is a CMOS device. The 4049 can handle up to 3.2 mA of current, so it will drive two standard TTL inputs (or fan-out of 2). The input voltages applied to the 4049 device can swing to +15VDC, regardless of the voltage applied to pin number 1.

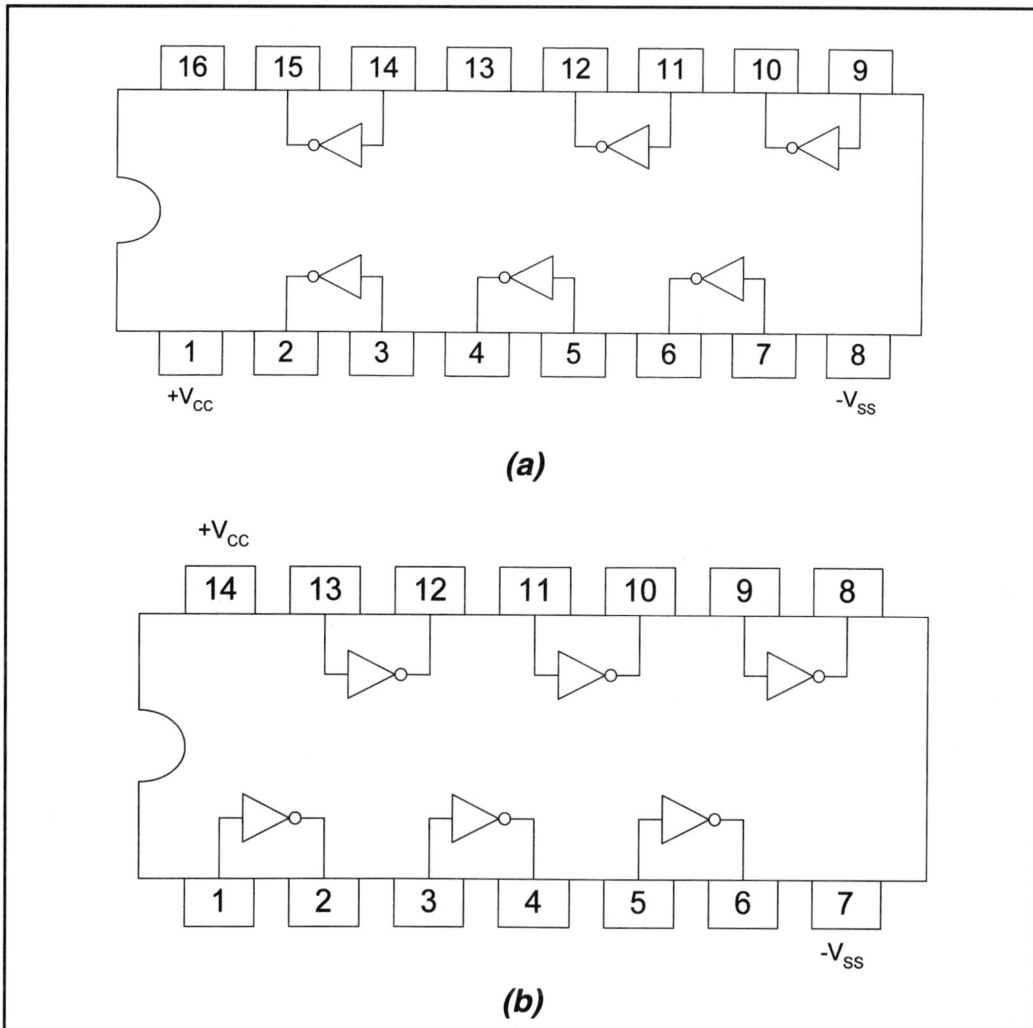

***Figure 5-3**. Two hex inverters.*

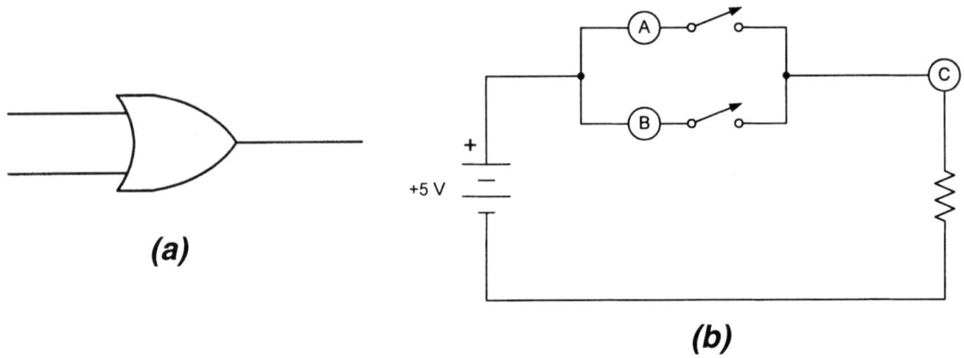

A	B	C
0	0	0
0	1	1
1	0	1
1	1	1

(c)

Figure 5-4. *OR gate: a. symbol; b. model circuit; c. truth table.*

Although some people call the 4069 device a "low-power" version of the 4049, it is not. The 4069 has different inputs and is not suitable for either direct interfacing with TTL logic or level translation of any type. The 4069 will interface directly with other CMOS devices.

The difference between the current drains of TTL and CMOS devices can be seen by comparing the 7404 and 4069 devices. At +5VDC, which is used by traditional TTL devices, the 4069 device requires only 0.5 mA, as opposed to 15 mA for the 7404. This is a difference of 3,000 percent! It is little wonder that designers have opted for the CMOS and related families wherever possible.

OR Gates

An OR gate produces a logical-1 (or "HIGH") output if any input is logical-1. *Figure 5-4a* shows the schematic circuit symbol for an OR gate, while *Figure 5-4b* shows an equivalent circuit made from ordinary switches. The model circuit produces the same sort of action. *Figure 5-4c* is the truth table for the OR gate.

Consider the circuit of *Figure 5-4b* in light of the definition of the OR gate given above. The voltage appearing at point "C" represents the output of the OR gate, while switches "A" and "B" represent the inputs. The switches' logic states are as follows:

Switch Condition	Input Logic Level	Output Condition
Open	0 (LOW)	LOW
Closed	1 (HIGH)	HIGH

By the definition of an OR gate, we would expect "C" to be HIGH when either input is HIGH. In other words, if either switch "A" OR switch "B" (denoted A OR B) is closed, then the output at "C" is HIGH. This is exactly the action found in the circuit of *Figure 5-4b*. Voltage C goes to +5VDC if either switch is closed and will be zero only when both switches are open. This condition is reflected in the truth table shown in *Figure 5-4c*: C is HIGH if A OR B is also HIGH.

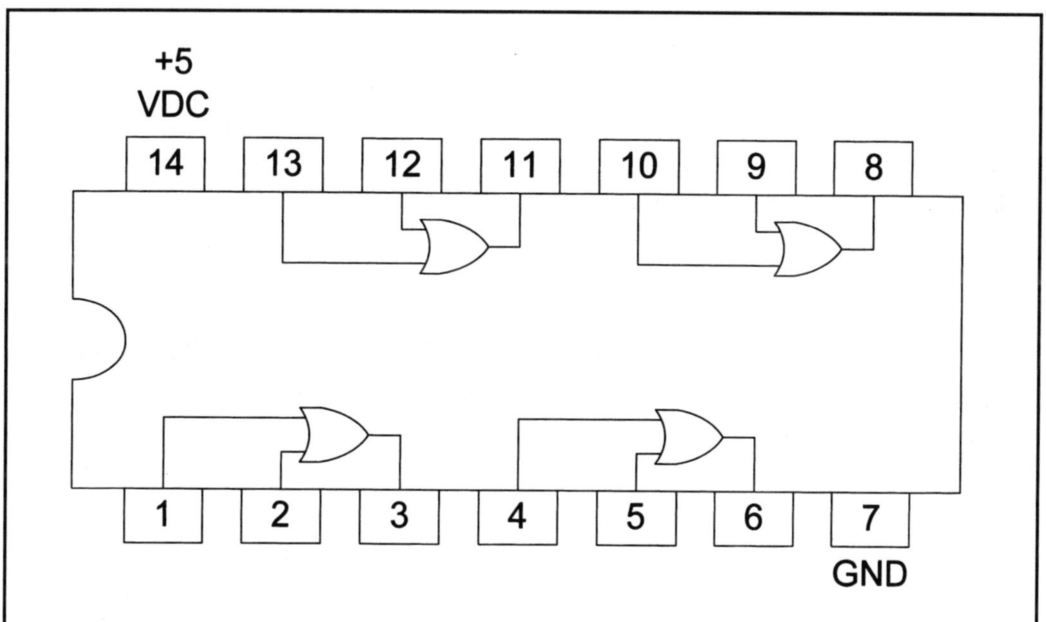

***Figure 5-5**. TTL quad OR gate.*

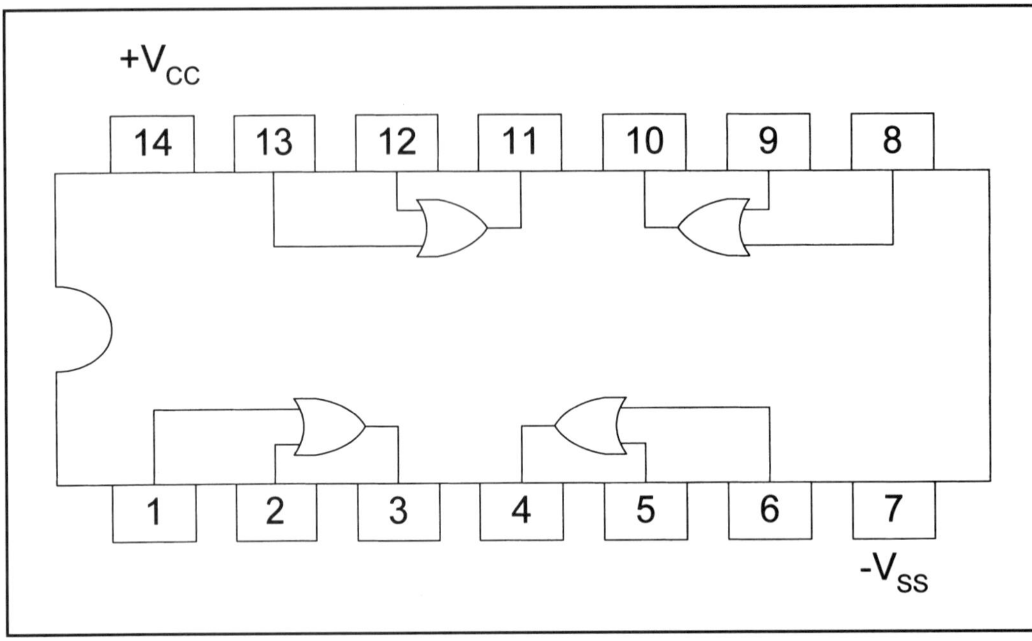

***Figure 5-6**. CMOS quad OR gate.*

Examples of TTL and CMOS OR gate ICs are shown in *Figure 5-5* and *Figure 5-6*, respectively. The type 7432 device is a quad, two-input OR gate. Note that the V+ and ground terminals are pins 14 and 7, respectively. Other than the power supply connections, all four gates are independent. These are the same DC power supply pinouts found on the 7404 device discussed earlier. The 7432 requires approximately 20 mA of current, although newer devices require considerably less.

The CMOS example, a 4071 device, is shown in *Figure 5-6*. This IC is also a quad, two-input OR gate. The current requirement is only 0.5 mA, so you can see the advantage that CMOS has over TTL (where speed is not a factor). The propagation delay time is 12 nanoseconds (ns) for TTL and 80 ns for CMOS.

AND Gates

The AND gate produces a HIGH output if and only if both inputs are HIGH, and a LOW output if *either* input is LOW. *Figure 5-7a* shows the circuit symbol for an AND gate, while *Figure 5-7b* shows an equivalent switch circuit that does the same job. *Figure 5-7c* shows the AND gate truth table. In this type of circuit, we obtain a

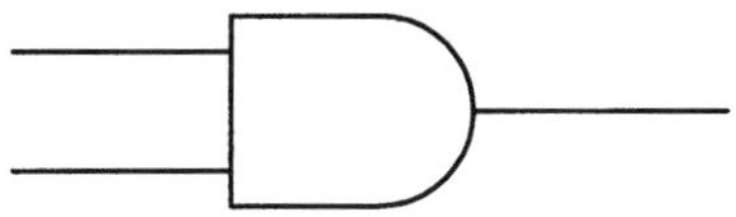

(a)

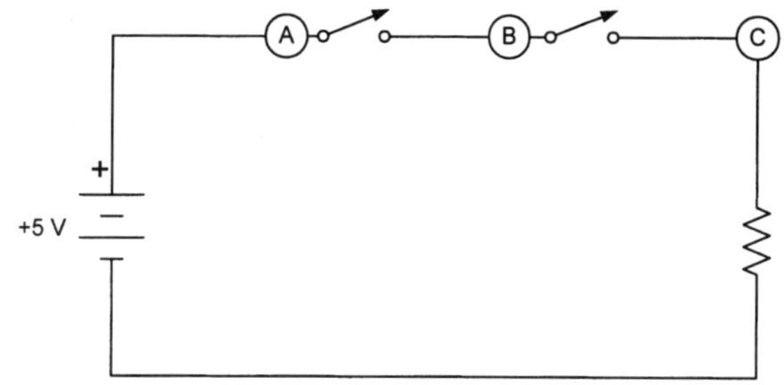

(b)

A	B	C
0	0	0
0	1	0
1	0	0
1	1	1

(c)

***Figure 5-7**. AND gate: a. symbol; b. model circuit; c. truth table.*

HIGH output only if A AND B are also HIGH. There are a few examples of TTL AND gates; but in the CMOS line, we find the 4073, 4081 and 4082 devices. The 4073 is a triple, three-input AND gate. The 4081 is a quad, two-input AND gate, and the 4082 is a dual, four-input device.

NAND Gates

A NAND gate can be viewed as an AND gate with an inverted output. In fact, you can simulate the NAND gate by connecting the input of an inverter (NOT gate) to the output of an AND gate. The designation "NAND" is derived from NOT-AND. We sometimes write the NAND expression in the forms:

$$\overline{A \text{ AND } B}$$

$$\overline{A \times B}$$

$$\overline{A \bullet B}$$

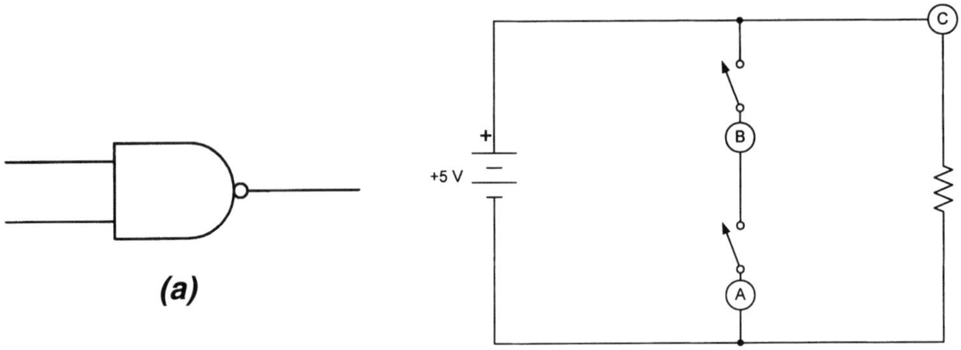

(a)

(b)

A	B	C
0	0	1
0	1	1
1	0	1
1	1	0

(c)

***Figure 5-8**. NAND gate: a. symbol; b. model circuit; c. truth table.*

Figure 5-8 shows the NAND gate. *Figure 5-8a* shows the circuit symbol used in schematic diagrams. *Figure 5-8b* is an equivalent switch circuit, and *Figure 5-8c* is the NAND gate truth table. The rules of operation for the NAND gate are:

1. A LOW on either input creates a HIGH output.
2. Both inputs must be HIGH for the output to be LOW.

We can see this same action in the equivalent circuit shown in *Figure 5-8b*. Recall our switch logic protocol: switch open is a LOW on the input, and switch closed is a HIGH on the input.

The voltage at point "C" will be HIGH if either switch A or switch B is open (either input is LOW). But if both switches are closed (both inputs are HIGH), then the load is shorted out and point "C" is LOW.

The NAND gate is one of the most popular gates used in digital electronic circuits. An example of a quad, two-input, TTL NAND gate is shown in *Figure 5-9*.

Recall that TTL is a current-sinking logic family, meaning that the output will sink (pass to ground) electrical current. The inputs of a TTL device form a 1.6 mA current source. Any traditional TTL input is considered HIGH when greater than +2.4 volts is applied, and must be held to less than 0.8 volts when LOW. This requirement means that any ground connection must have a resistance less than R = (0.8V)(0.0018A) = 440 ohms. If a higher ground resistance is encountered, or if a voltage between 0.8 volts and 2.4 volts is applied, then operation is undefined; all bets for proper operation are off.

A CMOS NAND gate IC, such as the old 4011 device, is shown in *Figure 5-10*. This device is also a quad two-input NAND gate. The 4011 requires only 0.4 mA per package (as opposed to 12 mA for TTL). Various combination circuits using NAND gates will be discussed in subsequent chapters.

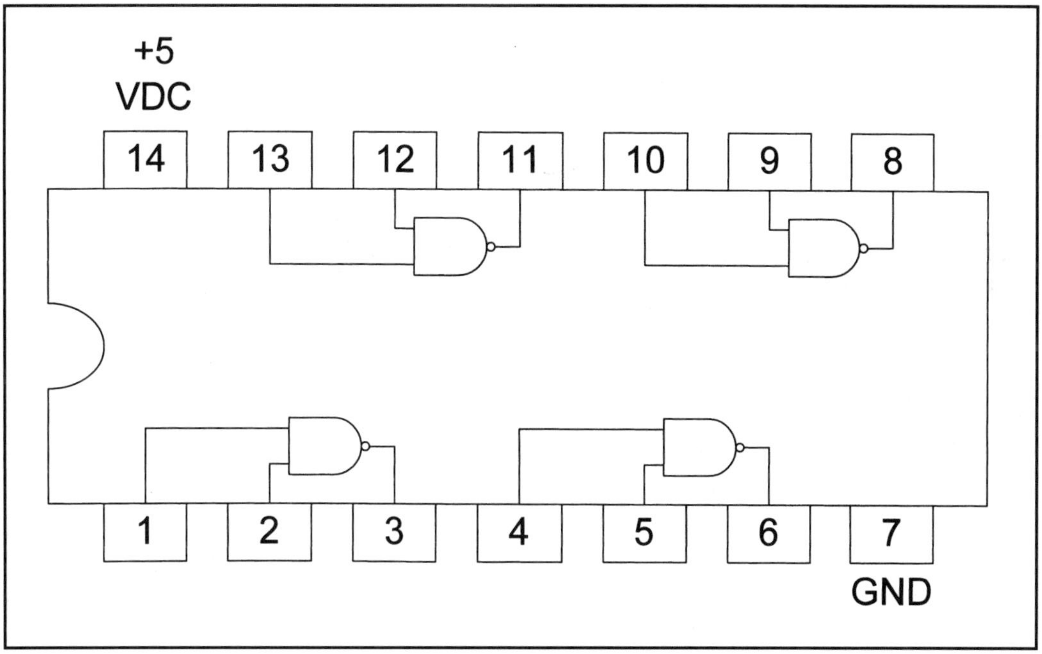

Figure 5-9. *TTL quad NAND gate.*

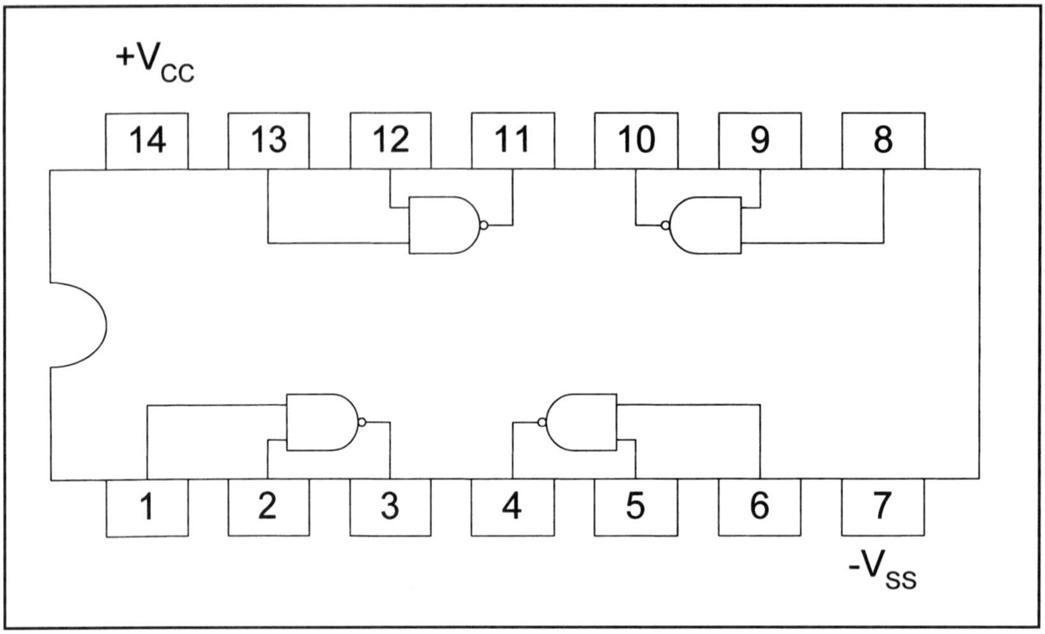

Figure 5-10. *CMOS quad NAND gate.*

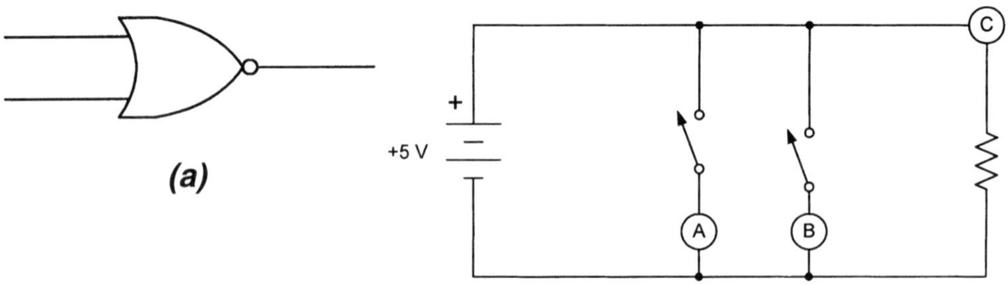

(a)

(b)

A	B	C
0	0	1
0	1	0
1	0	0
1	1	0

(c)

Figure 5-11. *NOR gate: a. symbol; b. model circuit; c. truth table.*

NOR Gates

A NOR gate is the same as an OR gate with an inverted output. The NOR designation is derived from NOT-OR, and its operation is described by *bar over A + B*. The circuit diagram symbol for the NOR gate is shown in *Figure 5-11a*; it is the OR gate with a circle on the output to signify the inversion. The equivalent circuit is shown in *Figure 5-11b* and the truth table is in *Figure 5-11c*. The rules governing the NOR gate are:

1. A HIGH on either input produces a LOW output.
2. Both inputs must be LOW to make the output HIGH.

We can see this action in the circuit of *Figure 5-11b*. If either switch A OR switch B is closed, then the output "C" will be HIGH.

Figure 5-12 shows an example of a TTL NOR gate. This device is not often used anymore, but serves to illustrate the operation. This device contains four two-input NOR gates that can be used independently of each other.

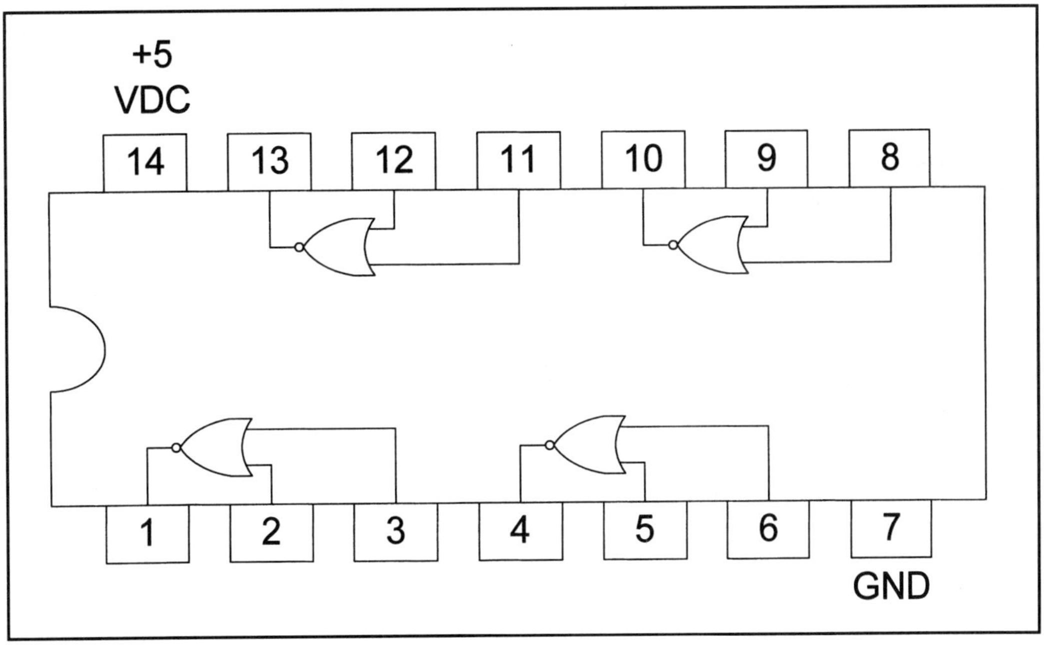

***Figure 5-12**. TTL quad NOR gate.*

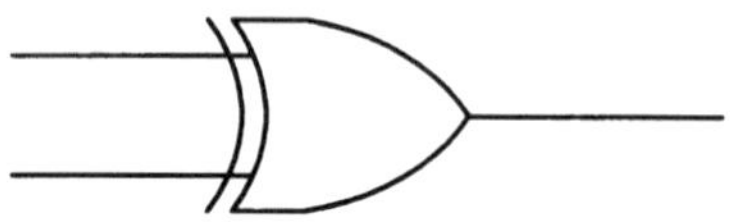

(a)

A	B	C
0	0	0
1	0	1
0	1	1
1	1	0

(b)

***Figure 5-13**. XOR gate: a. symbol; b. truth table.*

Exclusive-OR (XOR) Gates

The exclusive-OR, or XOR, gate is shown in *Figure 5-13*. From the truth table, we may infer the rules of operation governing the XOR gate:

1. The input is LOW either if both inputs are LOW or both inputs are HIGH (that is, both inputs the same).

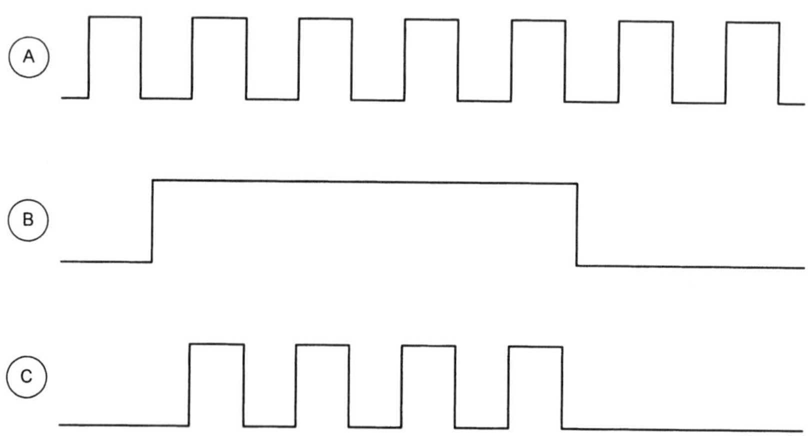

***Figure 5-14**. Timing diagram for an AND gate.*

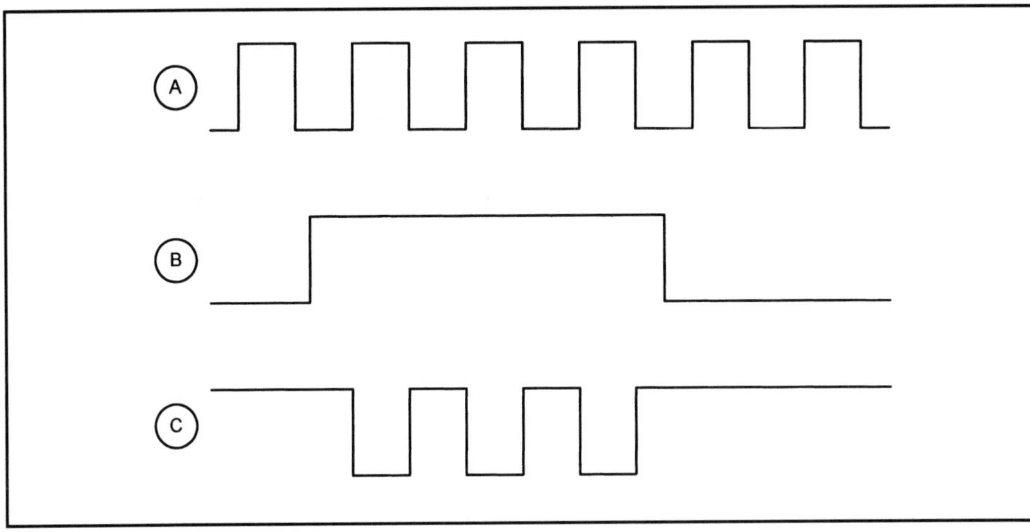

***Figure 5-15**. Timing diagram for a NAND gate.*

2. The output is HIGH if either input is HIGH.

In other words, a HIGH on either input alone will produce a HIGH output, but a HIGH on both inputs simultaneously produces a LOW output. Any time we find all inputs of an XOR gate in the same condition (HIGH or LOW), the output will be LOW.

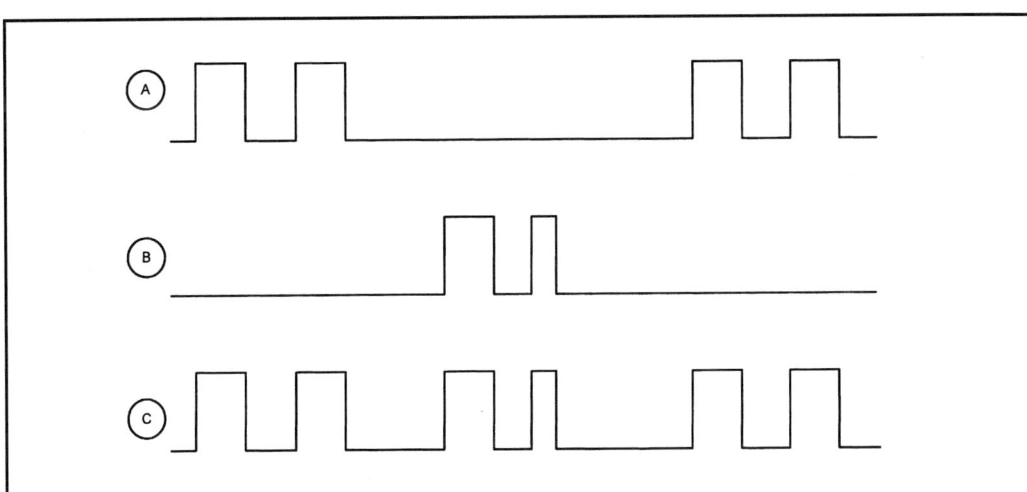

***Figure 5-16**. Timing diagram for an OR gate.*

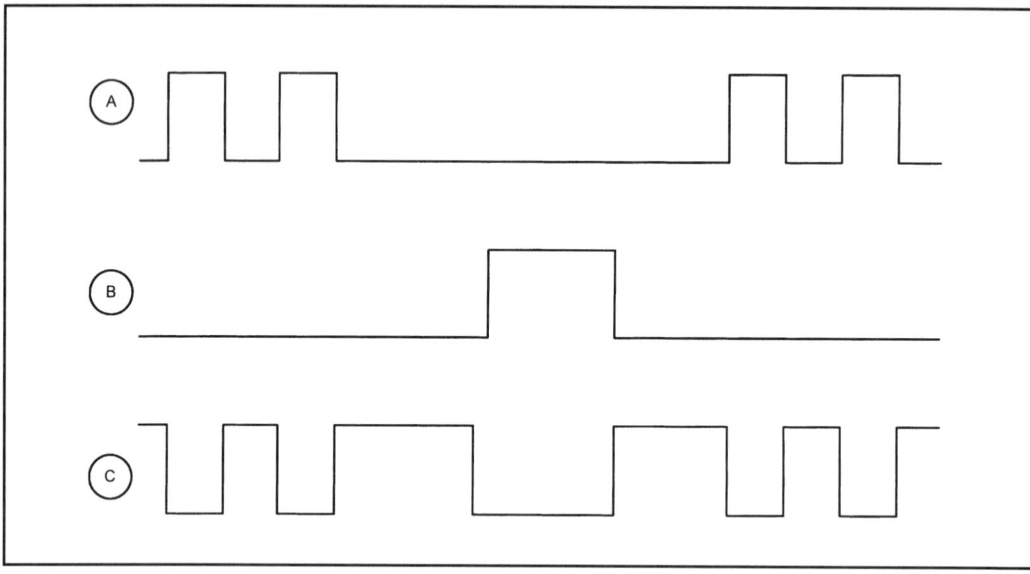

***Figure 5-17**. Timing diagram for a NOR gate.*

Summary of Gate Actions

We have thus far discussed five basic gates (AND, OR, NAND, NOR and XOR) and the inverter (NOT gate). From these basic forms of gate, all digital functions can be constructed. Even the largest microprocessor chip is little more than a mammoth array of gates variously configured. If we want to split hairs even further, we can say that all functions can be generated by the AND, OR, XOR and NOT gates.

Figure 5-14 through *Figure 5-19* show a series of waveform timing diagrams that summarize the actions of these basic gates. Such diagrams are frequently used in digital electronics as a graphical means for describing dynamic circuit actions. Indeed, the dynamic view is sometimes the only way to characterize performance.

In the timing diagrams, we have followed the same denotation as before: A and B are inputs, and C is the output. Note that these diagrams are for two-input devices (except, of course, for the inverter). But the basic concepts can be extended to devices with three or more inputs as well.

In *Figure 5-14*, we see the operation of the AND gate. In this case, a train of square waves is applied to input A, while a single pulse is applied to input B. Note that

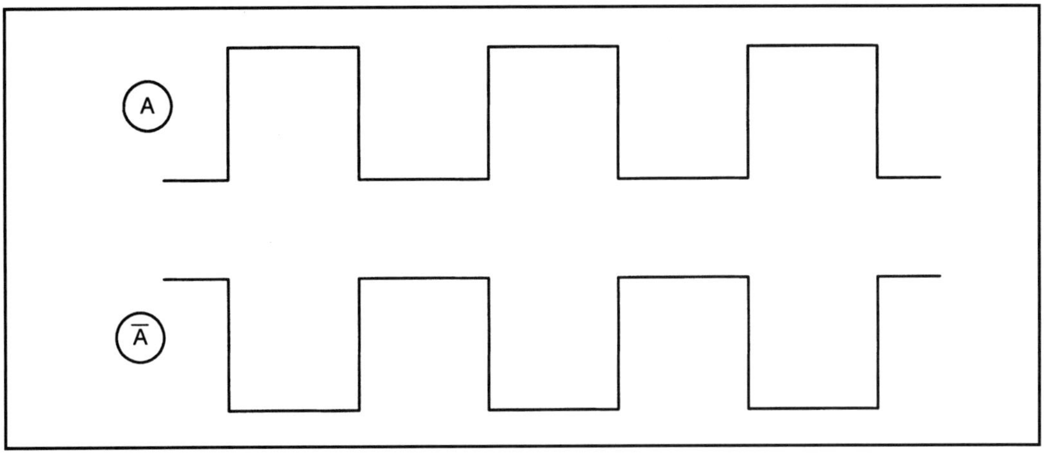

Figure 5-18. *Timing diagram for a NOT gate.*

output C remains LOW until input B goes HIGH. This is due to the requirement that both inputs must be HIGH before the output can be HIGH. Once input B is HIGH, because of the same rule, the output will follow input A.

A timing diagram for the NAND gate is shown in *Figure 5-15*, for the OR gate in *Figure 5-16*, NOR gate in *Figure 5-17*, NOT gate in *Figure 5-18* and the XOR in *Figure 5-19*. You are encouraged to study these waveforms to better understand how the gates operate.

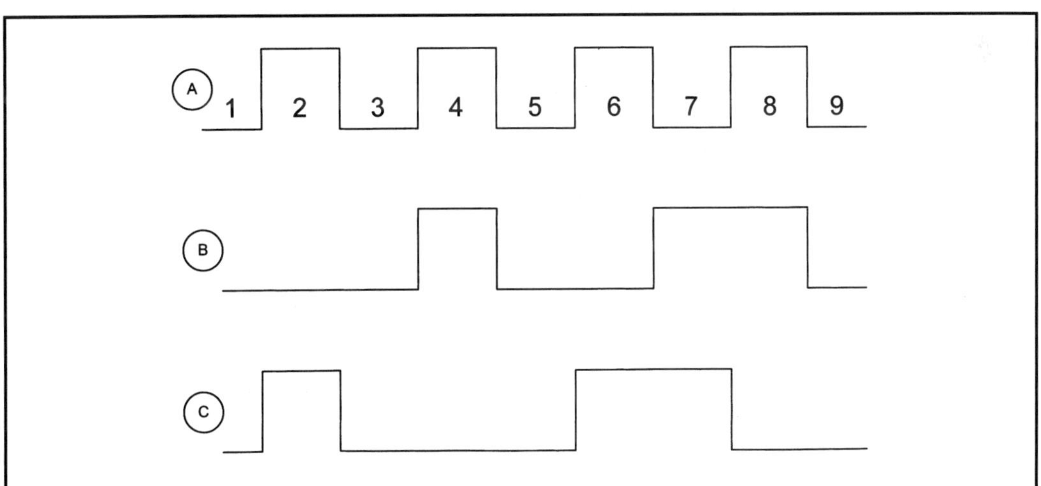

Figure 5-19. *Timing diagram for an XOR gate.*

Chapter 6

Unclocked Flip-Flops

All computer and digital logic functions can be made from combinations of AND, OR, and NOT gates. Even the NAND and NOR functions are only variants of NOT-OR and NOT-AND configurations. One of the problems with gates is that they are transient circuits; that is, they are incapable of storing information for even a short period of time.

Flip-flops are circuits that can store single bits of information. Many forms of solid-state computer memory devices are little more than arrays of flip-flops organized in a manner that allows for the storage of digital words in a computer.

Most flip-flops are bistable devices; that is, they have two stable output states. The flip-flop is not particularly concerned about which state is in existence at any given time. Being bistable, it is happy in either state. If the output is labeled Q, we find that either Q = HIGH or Q = LOW is an acceptable state, and will be stable. Some flip-flips are equipped with two complementary outputs, labeled Q and NOT-Q (or bar-Q).

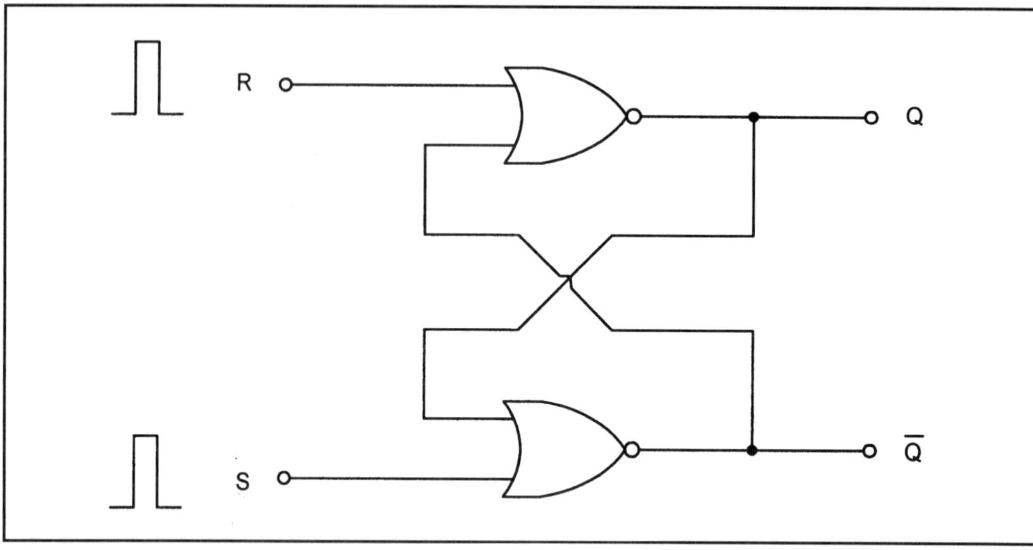

***Figure 6-1**. NOR gate R-S flip-flop.*

R-S Flip-Flops

One of the simplest flip-flops is the reset-set (R-S) flip-flop. These circuits can be made from either NAND or NOR gates, although the performance attributes are different for the two different forms. The NOR-logic configuration is shown in *Figure 6-1*, while the occasionally used circuit symbol is in *Figure 6-2*.

Note that the circuit has two outputs, Q and NOT-Q. This is an example of a complementary output flip-flop. These outputs always have the opposite state. If Q = HIGH then NOT-Q = LOW, and vice versa.

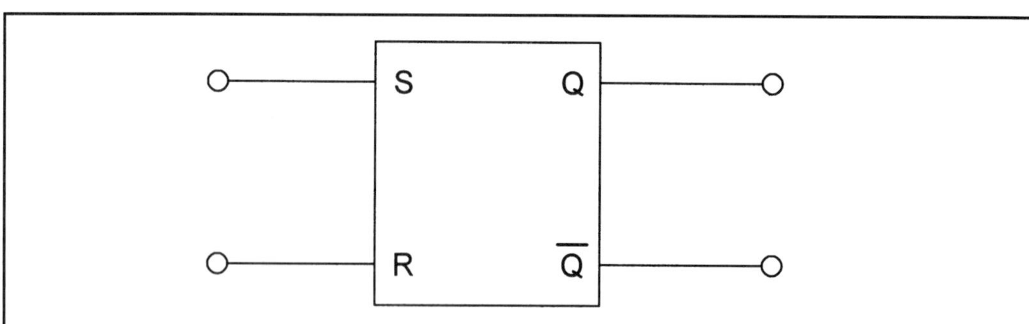

***Figure 6-2**. NOR R-S F-F symbol.*

In the NOR-logic R-S flip-flop, the output changes state when an appropriate input is momentarily brought HIGH from its resting LOW position. Only a brief pulse at the input is needed to effect the change. The rules governing the operation of the NOR-logic R-S flip-flop are summarized in the truth table in *Table 6-1*. These rules are also summarized below:

S	R	Q	NOT-Q
0	0	N.C.	N.C.
1	0	1	0
0	1	0	1
1	1	N.C.	N.C.

***Table 6-1**. The rules governing the operation of the NOR-logic R-S flip-flop.*

1. When both S and R inputs are LOW, no change occurs in the output.

2. If both S and R are simultaneously brought HIGH, this is a disallowed state. This input condition should be avoided.

3. Bringing the S input momentarily HIGH will set the output — makes Q = HIGH and NOT-Q = LOW.

4. Bringing the R input momentarily HIGH resets the flip-flop — makes Q = LOW and NOT-Q = HIGH.

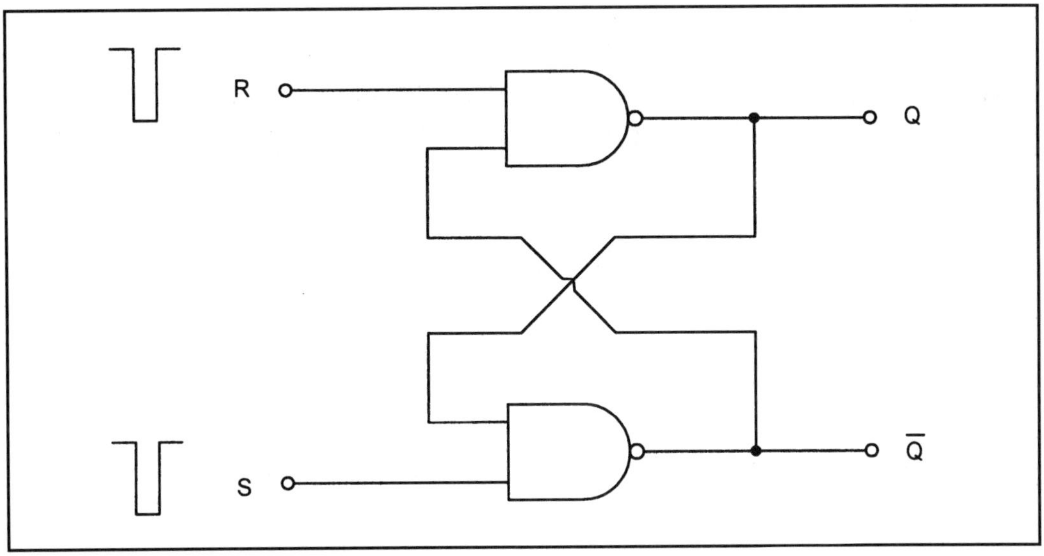

***Figure 6-3**. NAND gate R-S flip-flop.*

The basic NOR gate R-S flip-flop can be constructed from either TTL, CMOS or other logic family NOR gate arrays.

The NAND-logic R-S flip-flop uses inverted logic — the output state changes are caused by bringing the appropriate input LOW. An example of a NAND-logic R-S flip-flop is shown in *Figure 6-3* and *Figure 6-4*.

Because the inputs are active-LOW, this circuit is sometimes called a NOT-R-S flip-flop. In this text, however, it is more convenient to simply refer to both NAND-logic and NOR-logic circuits as simply R-S flip-flops. The NAND/NOR distinction is made in the text where necessary, or in the schematic.

The output conditions of the NAND-logic R-S flip-flop are shown in *Table 6-2*, and are summarized below:

1. If the S and R inputs are simultaneously LOW, then a disallowed state exists. This state must be avoided.

2. If both S and R inputs are made simultaneously HIGH, then no change in the inputs will occur.

3. If the S input is momentarily brought LOW, while the R input remains HIGH, then Q = HIGH and the NOT-Q = LOW. The flip-flop is in the "set" state.

4. If the R input is momentarily brought LOW, while the S input remains HIGH, then the flip-flop is reset — Q = LOW and NOT-Q = HIGH.

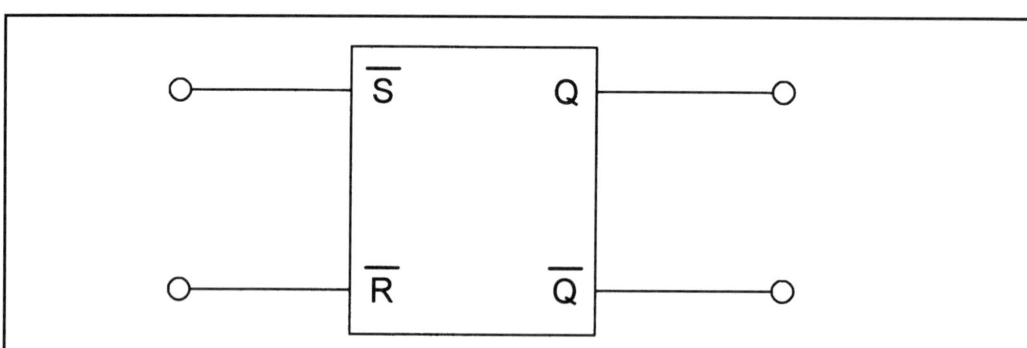

***Figure 6-4**. NAND R-S F-F symbol.*

The R-S flip-flop is used frequently in applications where a pulse applied to the S or R input sets a condition, and then a subsequent pulse is used to reset the circuit to its initial condition.

S	R	Q	NOT-Q
0	0	Disallowed	
1	0	1	0
0	1	0	1
1	1	Disallowed	

***Table 6-2**. The output conditions of the NAND-logic R-S flip-flop.*

Chapter 7

Clocked R-S Flip-Flops

The operation of the two previous R-S flip-flops, the NAND-logic and NOR-logic, was unconditional. The output state changed immediately when an appropriate input pulse was received. This is called *asynchronous operation.*

The clocked R-S flip-flop in *Figure 7-1* is able to operate *synchronously*; that is, the output will change state only when the input pulse coincides with a *clock* pulse. This behavior is obtained by adding a pair of NAND gates to the NAND-logic R-S flip-flop circuit.

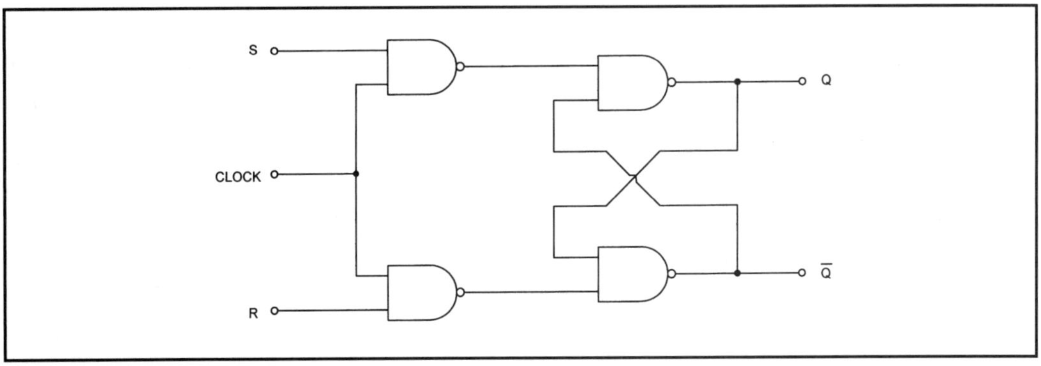

Figure 7-1. *Clocked R-S flip-flop.*

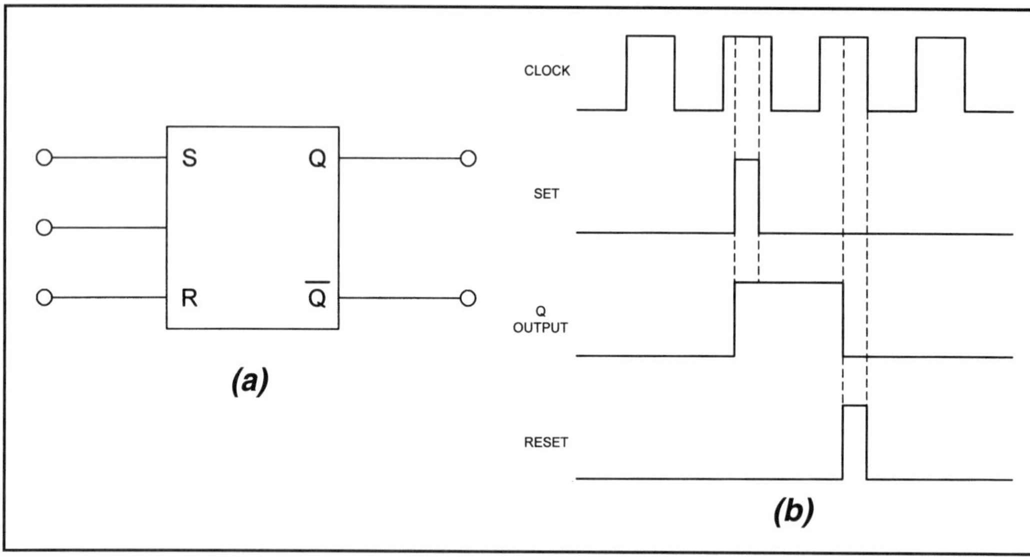

Figure 7-2. a. Clocked R-S F-F circuit symbol; b. timing diagram.

The commonly used schematic diagram circuit symbol for the clocked R-S flip-flop is shown in *Figure 7-2a*, while the timing diagram is in *Figure 7-2b*. The clock circuit produces a chain of square waves at the C input (also called CLK in some sources). Note that the Q output does not change state immediately when the S input pulse goes HIGH. The output waits until the clock input is also HIGH. Similarly, the output is not reset until the clock pulse and R input pulses are coincident. Because of this feature, the literature for some scientific instruments calls the clocked R-S flip-flop a *coincidence detector*.

Master-Slave Flip-Flops

It is often difficult, or impossible, to transfer data through a circuit in an orderly manner. We often get into high-speed digital analogies of the old-fashioned "relay race" problem. Recall the type of problem in circuits where two electromechanical relay switches are supposed to close simultaneously. If one relay is a little sluggish, or the other is a little snappier than the other (or both), then there is a brief instant where one is open and the other is still closed, or vice versa. This condition often produces unpredictable or even disastrous results. The same action exists in digital circuits where the culprit is differences in device *propagation time* — the time required for a pulse to travel through the circuit to effect an output change.

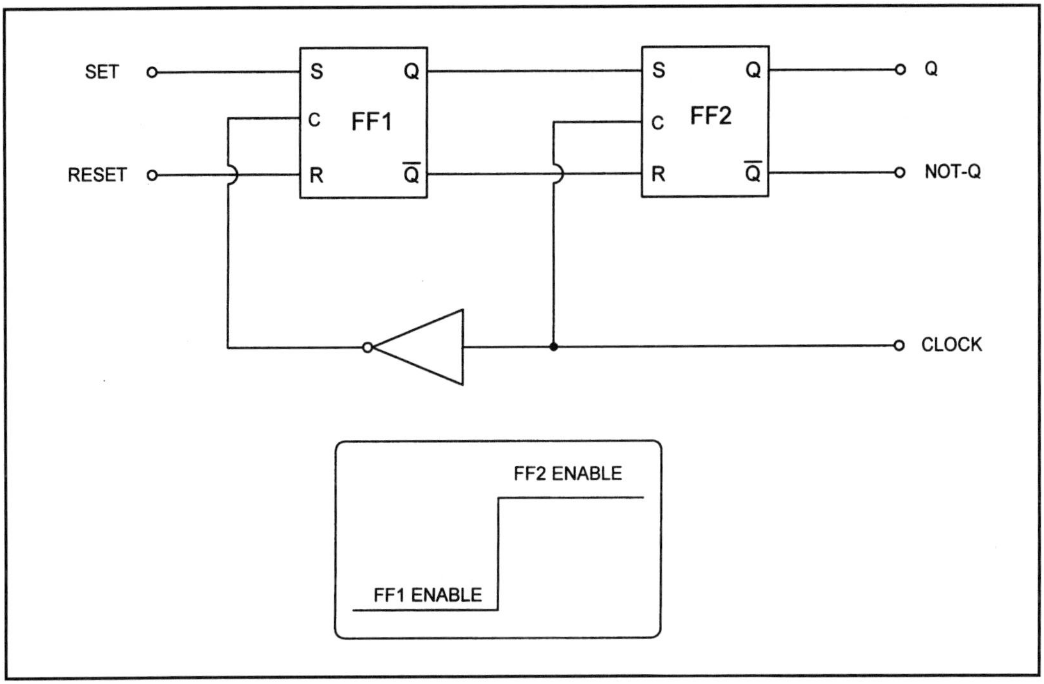

***Figure 7-3**. Master-slave F-F.*

Note that two clocked R-S flip-flops and an inverter can be used to synchronize data transfer. The circuit, which is shown in *Figure 7-3*, is called a *master-slave flip-flop*, also sometimes called the RST flip-flop.

In *Figure 7-3*, the inputs to FF1 are the inputs to the circuit as a whole, or the inputs to the master-slave FF. The outputs of the master-slave FF are the outputs of flip-flop FF2. Also, the clock affects FF2 directly, but must be inverted before being applied to FF1. Both flip-flops only go active when their respective clock inputs are HIGH.

A timing diagram for this circuit is shown in *Figure 7-4*. Recall from above that FF1 is enabled when the overall clock pulse is LOW, and FF2 is enabled when it is HIGH (because of the inverter action).

When the pulse is applied to the S input, nothing happens at Q1 (the Q output of FF1) until the clock pulse drops LOW again. At that time, Q1 snaps HIGH, thereby making the S input of FF2 HIGH also. But at this time the clock pulse is LOW, so no

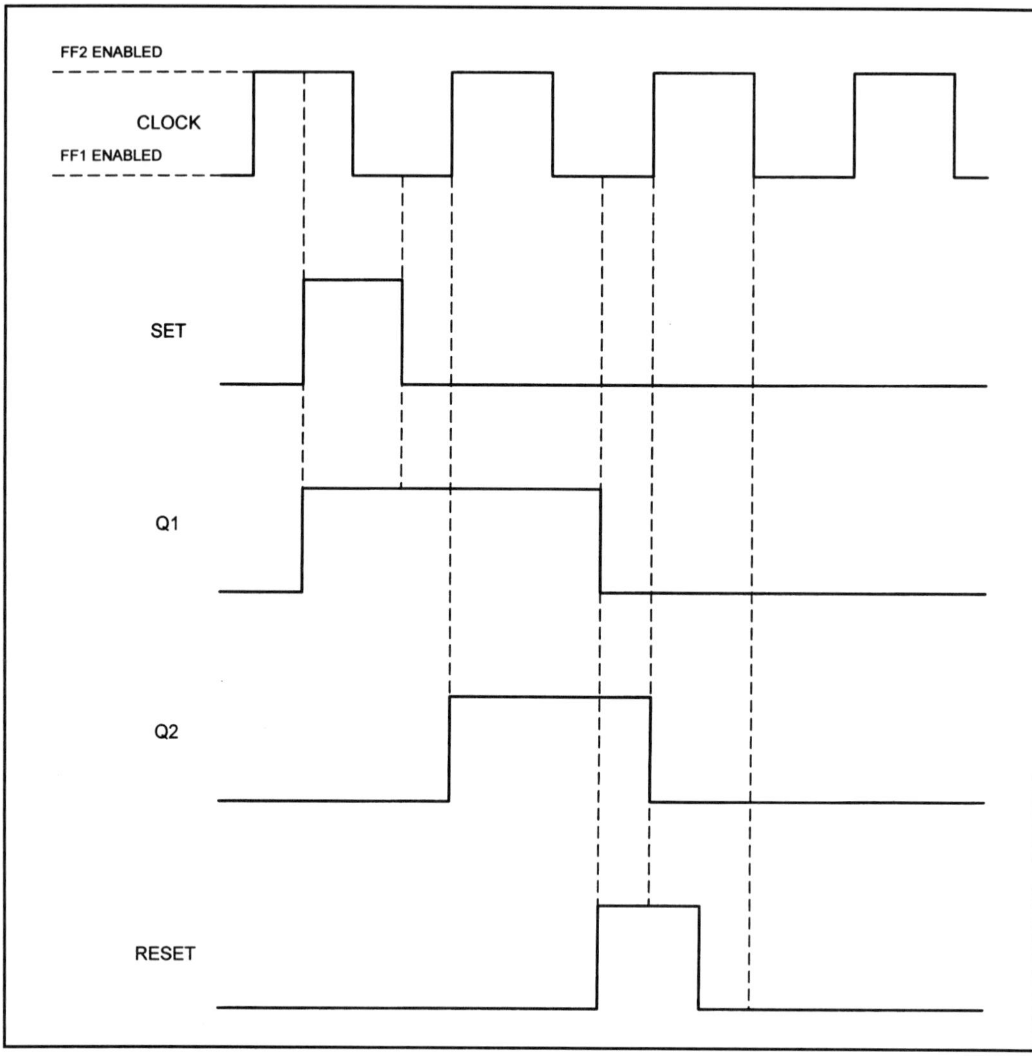

***Figure 7-4**. Timing diagram.*

change occurs at the FF2 output terminal. When the next clock pulse arrives, however, the FF2 clock input goes HIGH (Q1 and S2 are still HIGH), so the Q output of FF2 goes HIGH.

Similarly, the reset pulse arrives when the clock is LOW, so the Q1 output immediately drops LOW. The Q output of FF2, however, remains in the HIGH state until the next HIGH transition of the clock input.

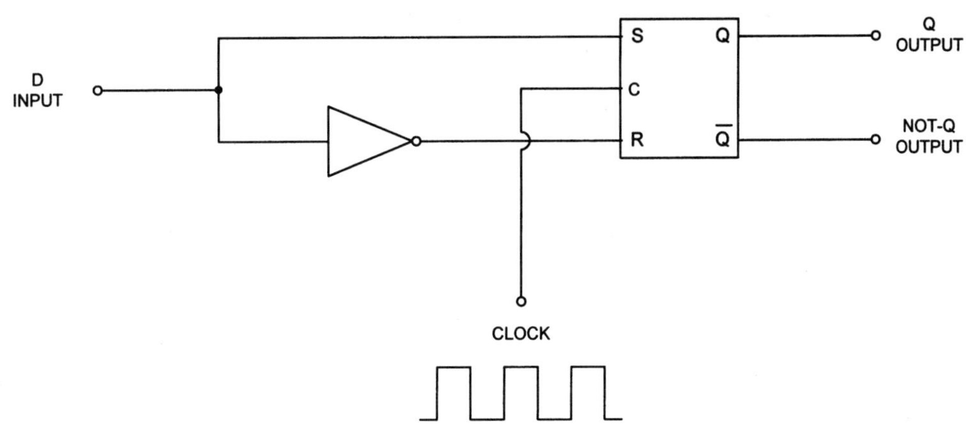

Figure 7-5*. Type-D flip-flop.*

FF1 is considered the master flip-flop, while FF2 is the slave flip-flop. The action at FF1 is given time to settle before the changes can be reflected in the output of FF2. This feature provides an orderly transfer of data between input and output.

Type-D Flip-Flops

A type-D flip-flop is a modified RST flip-flop that has only one input (labeled D for "data"). See *Figure 7-5.* The type-D flip-flop will transfer data (HIGH or LOW) from the D input to the Q output only when the clock terminal is HIGH. The following rules govern the behavior of the type-D flip-flop:

1. When the clock input goes HIGH, the data (HIGH or LOW state) present on the D input is transferred to the Q output.

2. If the clock input remains HIGH, then the Q output will follow changes in the data at the D input.

3. If the clock input remains LOW, then the Q output will retain the last data that was present on the D input at the instant the clock dropped LOW.

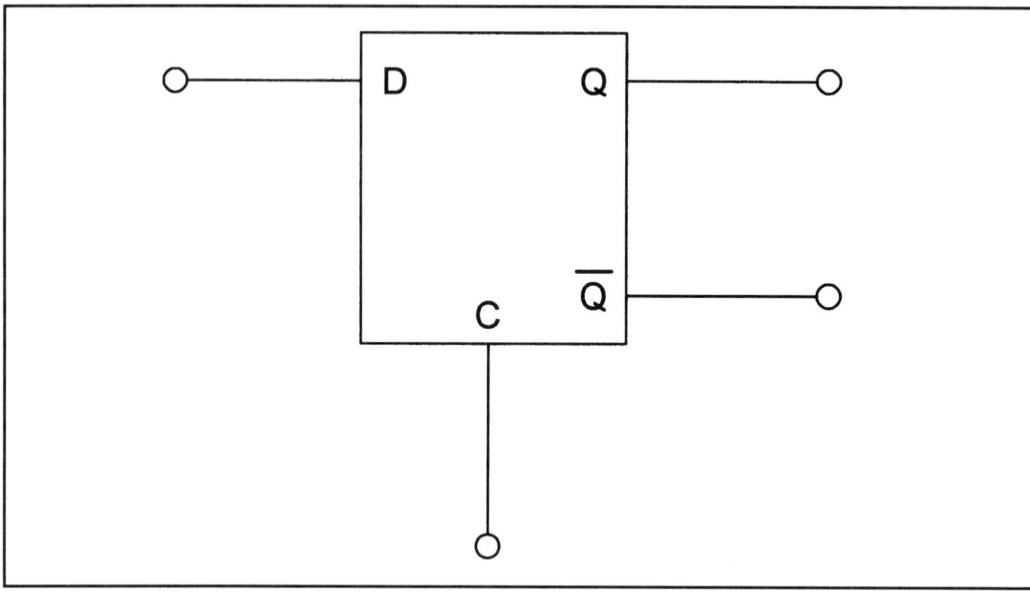

Figure 7-6. *Type-D F-F symbol.*

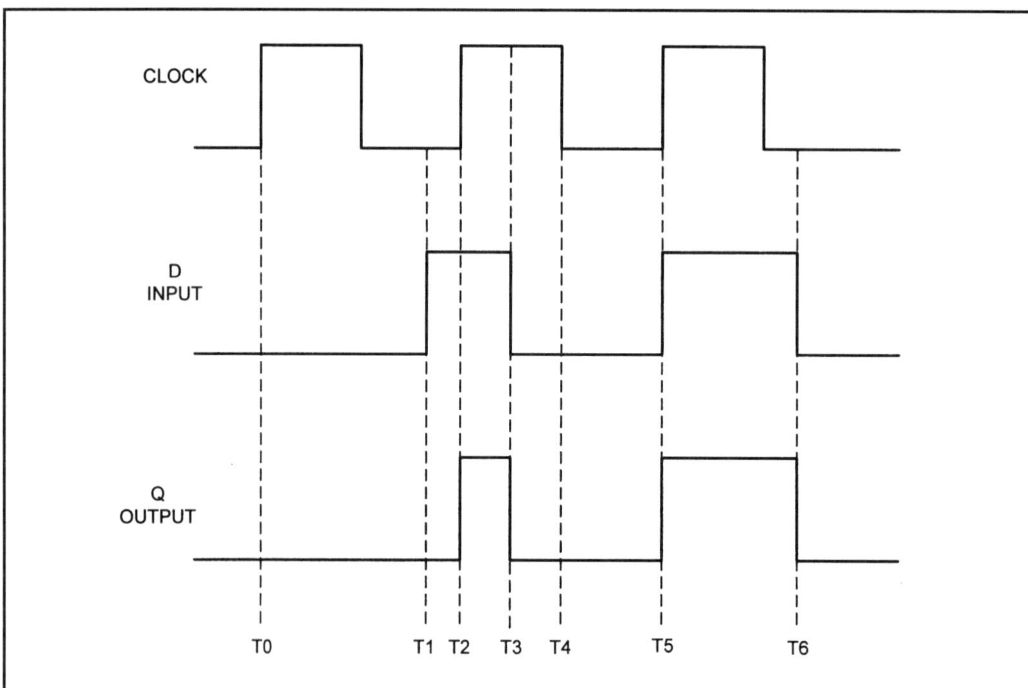

Figure 7-7. *Type-D timing diagram clocked mode.*

Because of the behavior presented in the above rules, the type-D flip-flop is sometimes called a *data latch*, or simply *latch*. The circuit symbol for the type-D flip-flop is shown in *Figure 7-6*.

We can see these rules more graphically in *Figure 7-7* through *Figure 7-9*. Consider first *Figure 7-7*. Recall that the data on the D input will be transferred to the Q output only when the clock terminal is HIGH. At time T0 in *Figure 7-7*, the clock goes HIGH, and the D input is LOW, so the Q output is also LOW.

At time T1, the D input goes HIGH but cannot affect the Q output because the clock is LOW. The clock goes HIGH at T2, so the output will also goes HIGH. The output pulse exists only for the interval T2-T3 because the HIGH condition on the D input only coincides with that interval.

At time T5, both the D and clock inputs go HIGH, but note that the D line remains HIGH even after the clock pulse disappears. By the third rule, then, the output must remain HIGH after the clock pulse passes.

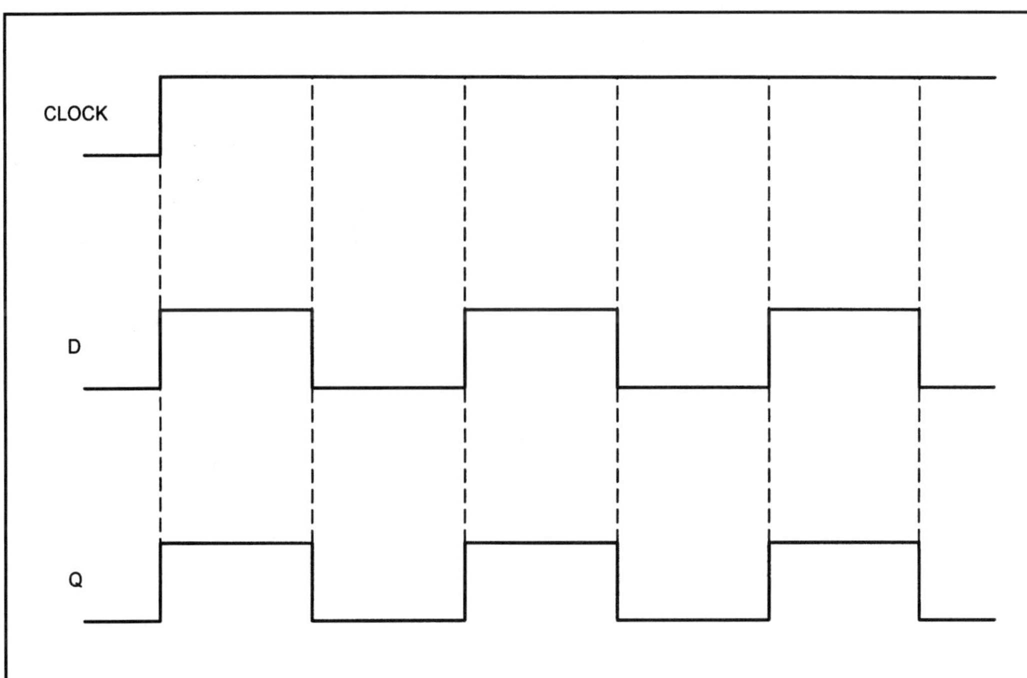

***Figure 7-8**. Type-D timing diagram direct mode.*

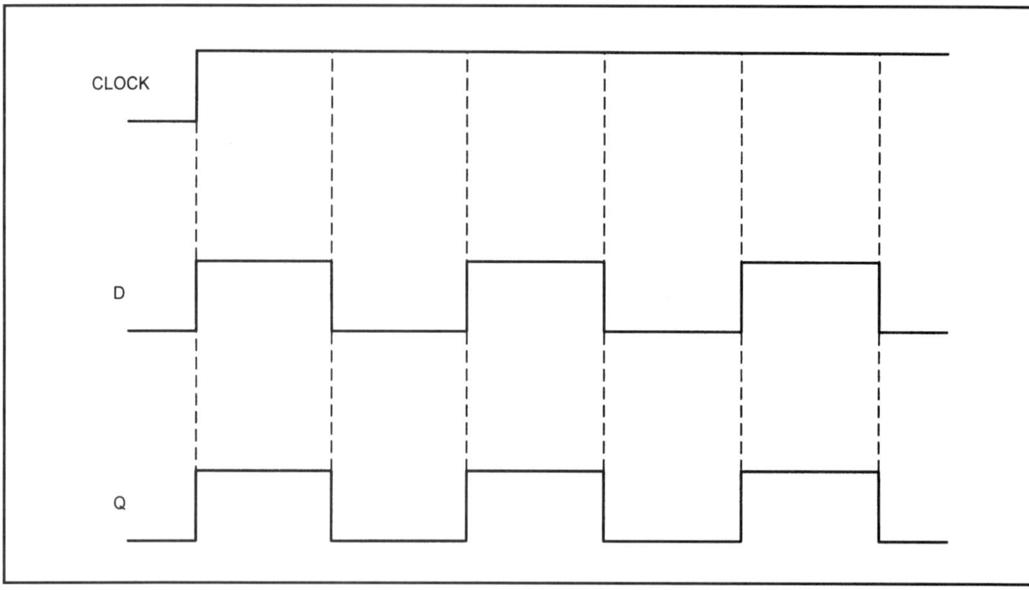

***Figure 7-9**. Timing diagram.*

Another situation is described by *Figure 7-8*. In this case, the clock line goes HIGH and remains HIGH. The Q output, therefore, is in an unlatched condition, so it will follow the data applied to the D input. Because the D input data is a square wave, the output will also be a square wave.

Still another condition is shown in *Figure 7-9*. Again, the D input data is a square wave or pulse train, but the block is not permanently HIGH in this example. At time T1, both the clock and the D input are HIGH, so the Q output is also HIGH. At time T2, the clock drops LOW, but since D is HIGH at that instant, the output will remain in the HIGH condition. Note that the D input can change at will, without affecting the output, as long as the clock input remains LOW. But at the time T5, the clock goes HIGH, and since the D line is LOW, the output goes LOW also. At time T6, however, the D line goes HIGH, while the clock is still HIGH, so the output goes HIGH.

Examples of Type-D Flip-Flops

Few people use individual gate chips to provide the flip-flop function anymore. The only reasonably common example where that occurs is in the R-S flip-flop. Even

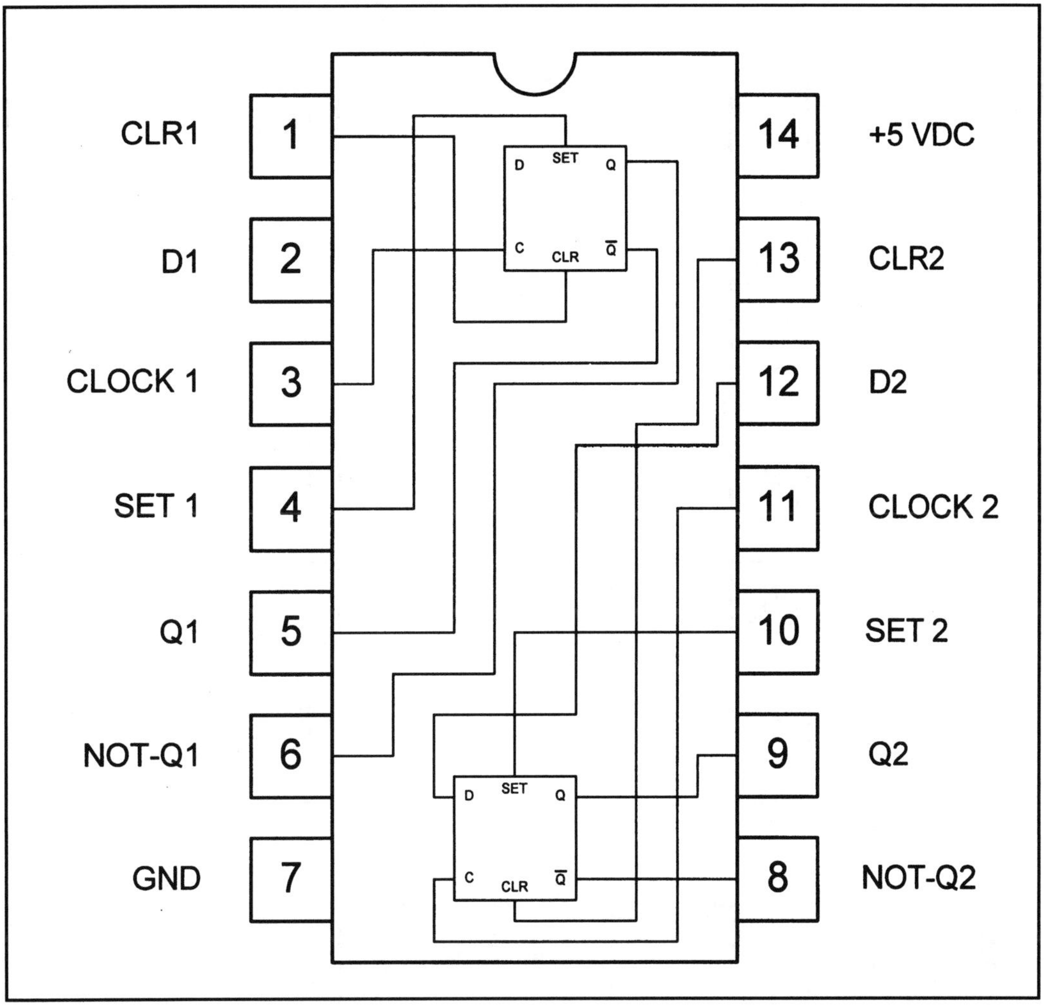

Figure 7-10. *Dual type-D F-F.*

those, however, are widely available in IC form, so they need not be individually constructed.

A commercial dual type-D flip-flop is shown in *Figure 7-10.* This particular type-D flip-flop has both *Set* and *Clear* inputs, in addition to the normal D and clock inputs. The device will obey the following rules:

1. Data on the D input is transferred to the Q output only on positive-going transitions of the clock pulse.

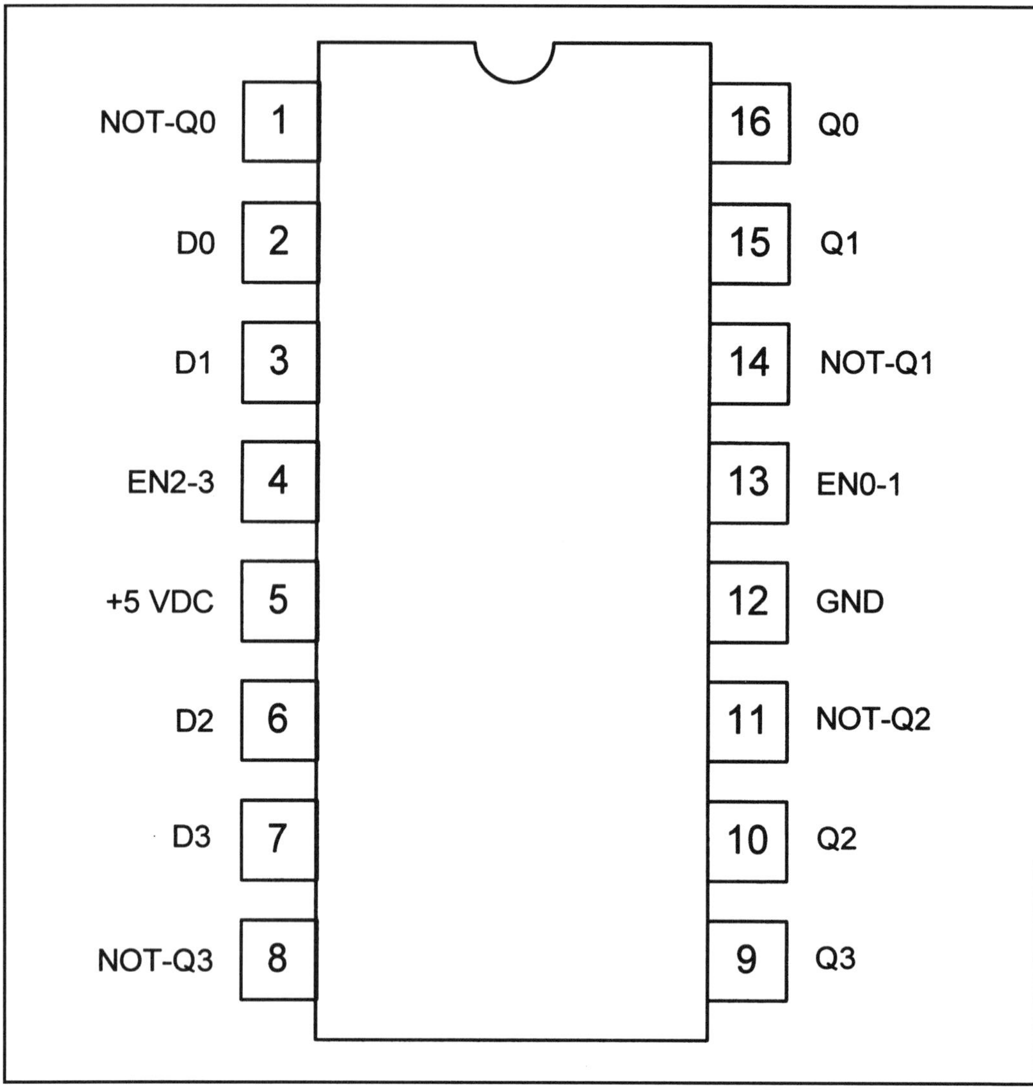

***Figure 7-11**. Pinouts.*

2. No changes in the data are reflected as output changes unless the clock is *going* HIGH (in the dynamic change state). This action is called *edge-triggering*.

3. If the *Set* input is grounded momentarily, then immediately Q = HIGH and NOT-Q = LOW. These conditions will remain in effect even if the *Set* input is returned to a HIGH state.

4. If the *Clear* input is grounded momentarily, then Q = LOW and NOT-Q = HIGH.

5. *Set* and *Clear* inputs must not be simultaneously grounded. If one or both of these inputs is not used, then it should be tied HIGH to prevent noise triggering.

6. The two flip-flops in the chip are totally independent of each other, except for the common DC power supply and ground connections.

The TTL 7475 device in *Figure 7-11* is a special case of a type-D flip-flop called a *quad latch*. This IC device is *level sensitive*, so the output will follow changes in the

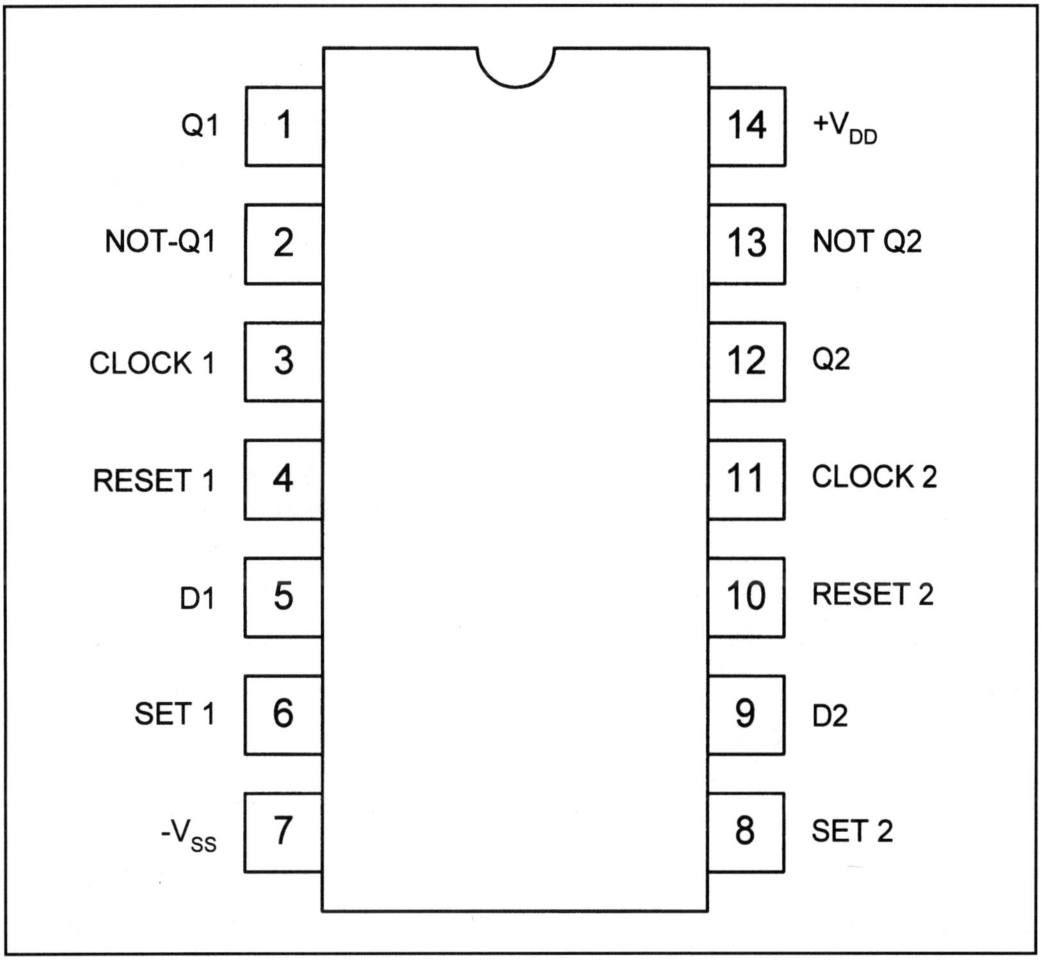

Figure 7-12*. Pinouts.*

input as long as the clock terminal is HIGH. This is more like the behavior expected of traditional type-D flip-flops, described earlier.

The 7475 is essentially obsolete, but is still used occasionally. It is often used to hold four bits of digital data — one bit is stored in each flip-flop. The pinouts for the 7475 device are shown in *Figure 7-11*. The clock inputs are labeled as *Enable* inputs. When the *Enable* inputs are HIGH, then data on the four D-inputs are transferred to the Q outputs. The 7475 is arranged in a 2 x 2 format, meaning that the *Enable* inputs of two flip-flops are tied together at a single package pin. The *Enable* inputs to the two remaining flip-flops are brought out to another package terminal.

One common place to find 7475s is in digital counter circuits, where they are used to hold the BCD data being displayed while the computer chip (which is often a 7490 device) updates the data. The display can be updated when the count is completed by momentarily bringing the *Enable* inputs HIGH. A single 7475 will store all four bits of the standard BCD word, so it can be used with a BCD counter and display decoder.

A device similar to the 7475 is the TTL 74100 device. This 24-pin device contains a pair of four-bit latches, and therefore has the ability to store up to eight bits of digital data.

A CMOS type-D flip-flop is shown in *Figure 7-12*. Like the 7474 device from the TTL line, this 4013 device is a dual type-D flip-flop. It can be used in either direct or clocked modes; that is, it has *Set* and *Clear* direct inputs as well as a D input and a clock input. The 4013 differs from the 7474, however, in that direct inputs become active when HIGH, rather than LOW. If unused, therefore, these inputs should be tied to ground to prevent noise triggering.

In the clocked mode, the 4013 device behaves much like the 7474 device. It is a positive-edge triggered flip-flop, so data is transferred from the D input to the Q output when the clock line goes HIGH. The action occurs on the positive-going transition of the clock pulse. In this respect, the 4013 and 7474 follow the same rules.

In the direct mode, the operation is similar to the 7474 but requires pulses of the opposite level. The rules for *Set* and *Clear* inputs are shown below:

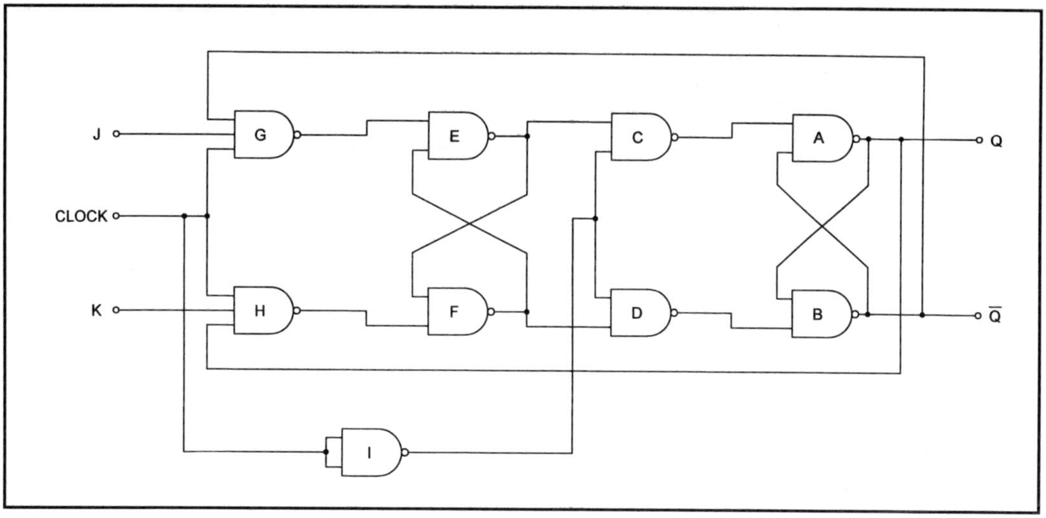

Figure 7-13. *J-K flip-flop.*

1. If the *Set* input is made HIGH, then Q = HIGH and NOT-Q = LOW.

2. If the *Clear* input is made HIGH, then Q = LOW and NOT-Q = HIGH.

3. Both *Set* and *Clear* must not be simultaneously HIGH or a disallowed state occurs.

J-K Flip-Flops

The J-K flip-flop is similar to the other clocked flip-flops, although certain critical differences exist that make the J-K flip-flop special. *Figure 7-13* shows a gate logic diagram of the J-K flip-flop, while the usual circuit symbol is shown in *Figure 7-14. Table 7-1* shows the truth table for the J-K. Like the type-D flip-flops discussed earlier, the J-K flip-flop is capable of operating in either direct or clocked modes.

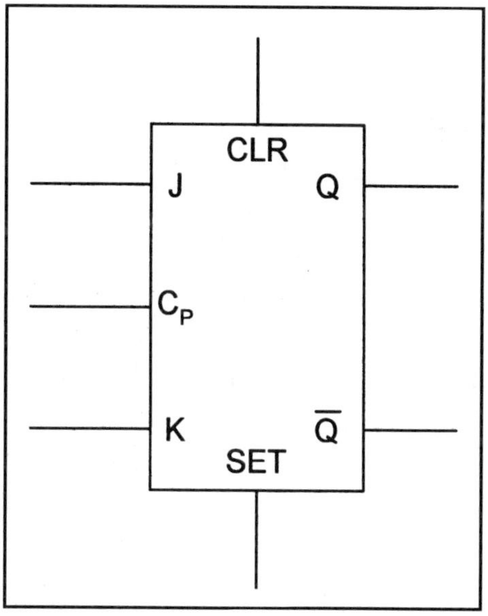

Figure 7-14. *J-K F-F symbol.*

Direct operation uses the *Set* and *Clear* inputs to force the Q and NOT-Q outputs to specific states, HIGH or LOW. The truth table for the TTL J-K flip-flops is shown in *Table 7-2*. These devices follow the rules below:

INITIAL CONDITIONS				FINAL CONDITIONS	
OUT		IN		OUT	
Q	NOT-Q	J	K	Q	NOT-Q
0	1	0	0	0	1
0	1	0	1	0	1
0	1	1	0	1	0
0	1	1	1	1	0
1	0	0	0	1	0
1	0	0	1	0	1
1	0	1	0	1	0
1	0	1	1	0	1

Table 7-1. *The truth table for the J-K.*

1. LOW conditions on both *Set* and *Clear* inputs simultaneously result in a forbidden, disallowed state.

2. Making the *Clear* input HIGH and the *Set* input LOW forces Q = HIGH.

3. Making the *Clear* input LOW and *Set* input HIGH forces Q = LOW.

4. Making both *Set* and *Clear* inputs HIGH causes the flip-flop to operate in the clocked mode.

Again, the CMOS version operates in a similar manner, but with opposite sense pulses. For example, the CMOS 4027 device is a dual J-K flip-flop that follows the rules below:

SET	CLEAR	Q
0	0	(DISALLOWED)
0	1	1
1	0	0
1	1	(NORMAL CLOCKED OPERATION

Table 7-2. *The truth table for the TTL J-K flip-flops.*

1. If both *Set* and *Clear* are LOW, then normal clocked operation occurs.

2. Making *Clear* LOW and *Set* HIGH forces Q = HIGH.

3. Making *Clear* HIGH and *Set* LOW forces Q = LOW.

SET	CLEAR	CLOCK	Q
0	0	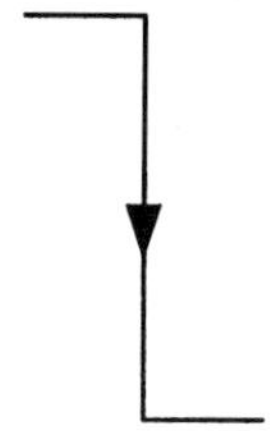	(NO CHANGE)
0	1		0
1	0		1
1	1		(GOES TO OPPOSITE)

***Table 7-3**. The operation of TTL J-K flip-flops in the blocked mode.*

4. Making both *Set* and *Clear* HIGH is a disallowed condition.

Clocked operation means that all changes occur synchronously with the clock pulse. The J-K flip-flop operates on the negative-going transition of the clock pulse, in distinct contrast to the behavior of the type-D devices discussed earlier.

Table 7-3 shows the operation of TTL J-K flip-flops in the blocked mode. The changes that are shown occur as illustrated on the negative transition of the clock pulse. The rules of operation are as follows:

1. If both J and K are LOW, then no change occurs in the output regardless of clock state.

2. If J is LOW and K is HIGH, then Q = LOW state is forced.

3. If J is HIGH and K is LOW, then Q = HIGH is forced.

4. If both J and K are HIGH, the Q output will go to the opposite of its present state. For example, if Q = HIGH, then Q will go LOW under this condition.

In the clocked mode, the CMOS 4027 device follows these same rules, except that the transitions occur on positive-going transitions of the clock pulse.

Chapter 8

Digital Arithmetic Circuits

In this chapter, you will learn about the *half-adder*, *adder* and *subtractor* circuits. These are known collectively as *arithmetic circuits*.

Adder and Half-Adder Circuits

There are two basic forms of adder circuit. These are called the *half-adder* and the *full-adder* (or simply *adder*). Of these, several different types of each exist. But before examining the various types of circuits, consider first the process of binary addition to learn about the requirements placed on any circuit that purports to be a binary adder. The rules for binary addition are reiterated here:

$0 + 0 = 0$
$0 + 1 = 1$
$1 + 0 = 1$
$1 + 1 = 0$ plus carry 1

Examine these rules and compare them with the truth table for the exclusive-OR (XOR) gate. The XOR gate obeys the following rules:

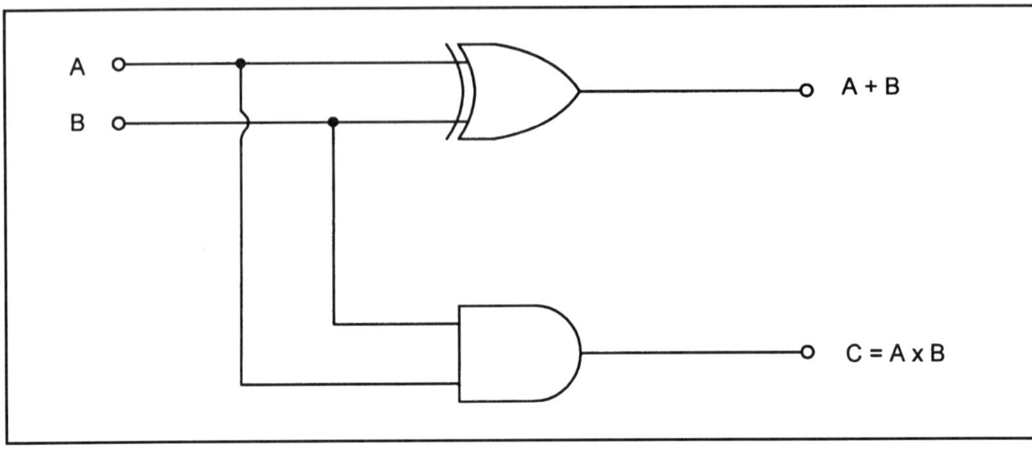

***Figure 8-1**. Half adder.*

1. If both inputs are zero (LOW, then the output is zero (LOW).

2. If both inputs are one (HIGH), then the output is zero (LOW).

3. If either, but not both, inputs are one (HIGH), then the output is one (HIGH).

The XOR gate truth table is, therefore:

A	B	OUTPUT
0	0	0
0	1	1
1	0	1
1	1	0

Note that the sole difference between the truth table for the XOR gate and the rules for binary arithmetic is that, in binary addition, a *carry-one* output is created if both A and B are one (HIGH). We can therefore create a binary addition circuit with an XOR gate and a means for generating the carry-one output.

Figure 8-1 shows a circuit for a half-adder. A half-adder is a circuit that generates two outputs: *Sum* (S = A + B) and *Carry* (C = A x B). (*Read*: S = A XOR B, and C = A AND B — the "x" symbol here denotes the logical operator "AND.") The exclusive-OR gate generates the sum output, while an AND gate generates the carry-

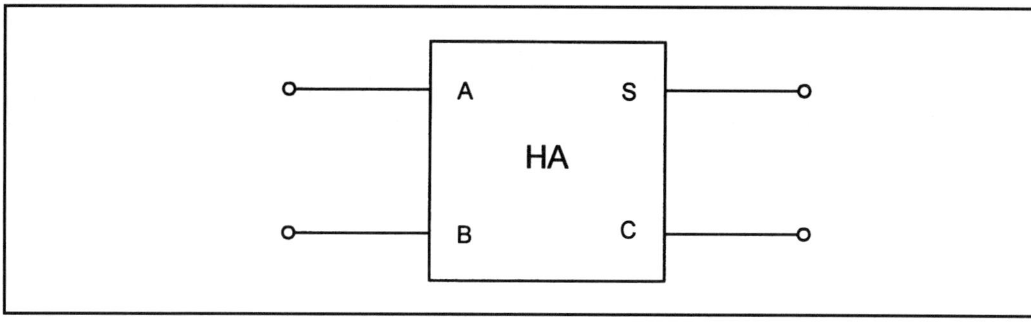

***Figure 8-2**. Half adder symbol.*

one output. In many cases, the logic symbol shown in *Figure 8-2* is used to denote the half-adder in logic diagrams, block diagrams and flow charts. *Table 8-1* shows a truth table for the half-adder.

A *full-adder*, or simply *adder* as it is usually called, is a circuit that will accept as a valid input the carry-one output from a lower order stage (another adder). Consider again the process of binary addition. Add two binary numbers:

```
        1  <— carry 1
        0  1
       +1  1
  1<—   0  0
```

In the first step, we added the two least significant digits. In this case, they are both ones, so the result is zero and a carry output 1 is generated. This carry 1 is added to the next most significant column. In the second step, we added a 0 and a 1 to obtain a 1 result. Next, we add the carry 1 from the previous step; so this is in effect a 1 + 1 operation. The result (written below the line) is a zero with a carry 1 generated. The result of this operation is a final result of 00 —> 1 (or zero with carry 1). The carry one can be ignored if there are no more significant digits.

HALF-ADDER TRUTH TABLE			
INPUTS		OUTPUTS	
A	B	S	C
0	0	0	0
0	1	1	0
1	0	1	0
1	1	0	1

***Table 8-1**. Half-adder truth table.*

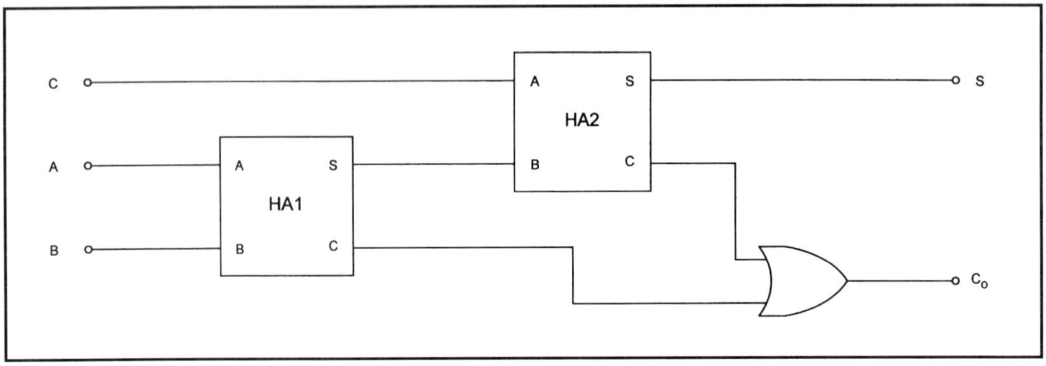

***Figure 8-3**. Full adder.*

The first operation in the above problem could be performed by a half-adder circuit because the least significant digit does not receive any carry bits from a lower order stage (there are no preceding stages to generate carries). All subsequent stages, however, must be full adder circuits so they can accommodate carries from previous stages. Otherwise, an erroneous output result could be obtained.

Figure 8-3 shows the logic diagram for a full-adder, which by definition will perform the following:

1. Add A and B (S = A XOR B).

2. Account for any carry-in C from lower order stages.

3. Produce a carry-out 1, C_o, if needed.

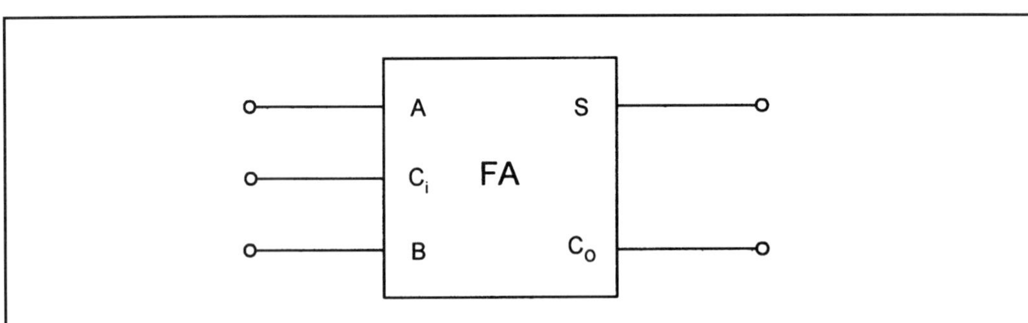

***Figure 8-4**. Full adder symbol.*

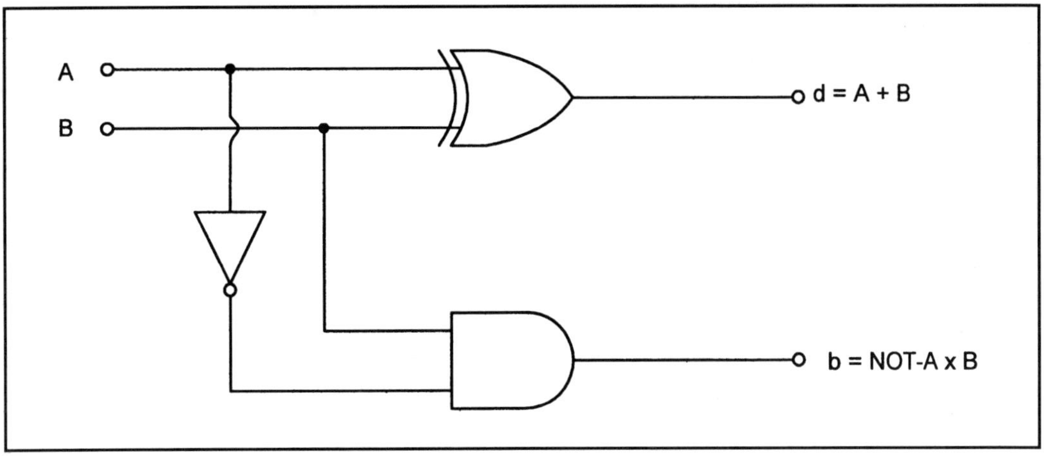

Figure 8-5. Adder.

The circuit of *Figure 8-3* consists of two half-adders and an OR gate to generate the carry output. Half-adder HA1 adds together the A and B inputs using the XOR function. This circuit will produce a carry-out 1 if needed. Half-adder HA2 adds together the results of HA1 (S') to produce the final output S = S' XOR C_i. A carry-out may also be generated by this stage. A commonly used logic symbol for the full-adder circuit is shown in *Figure 8-4*.

Subtractor Circuits

Addition and subtraction are very similar arithmetic operations. In fact, it is proper to view subtraction as the addition of a positive number and a negative number. It is not surprising, therefore, to find certain similarities between adders and subtractors used in digital circuits. In fact, the principal difference is the capability to account for a borrow-one from the next higher order digit.

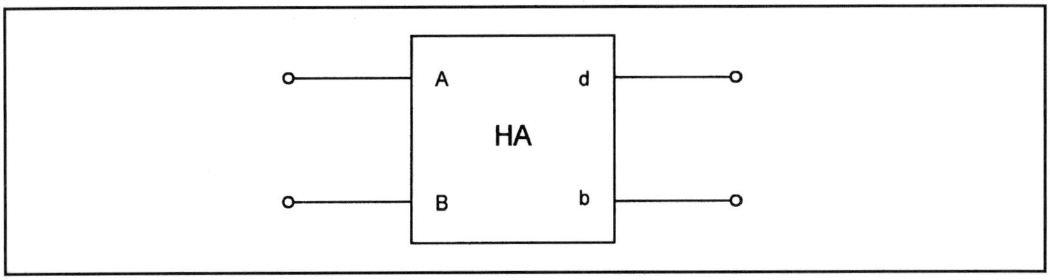

Figure 8-6. Circuit symbol.

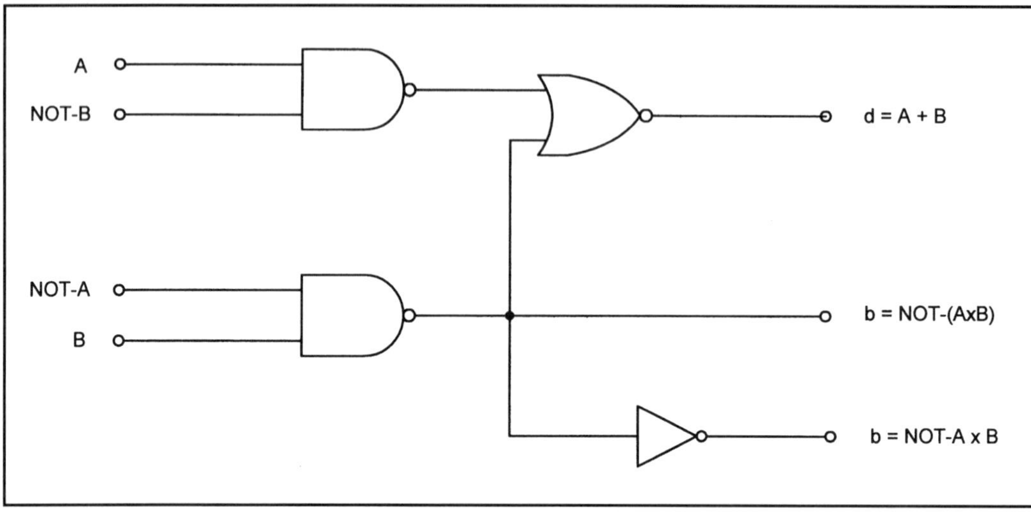

Figure 8-7*. Additional adder.*

Figure 8-5 shows a simple half-subtractor circuit which produces an output difference d = A XOR B, and a borrow-b of NOT-A and B. The logic symbol for the half-subtractor is shown in *Figure 8-6*.

An example of a half-subtractor circuit made entirely from standard TTL NAND, NOR and NOT gates is shown in *Figure 8-7*. This type of circuit is often found in digital equipment because the low cost of TTL IC devices makes it economical.

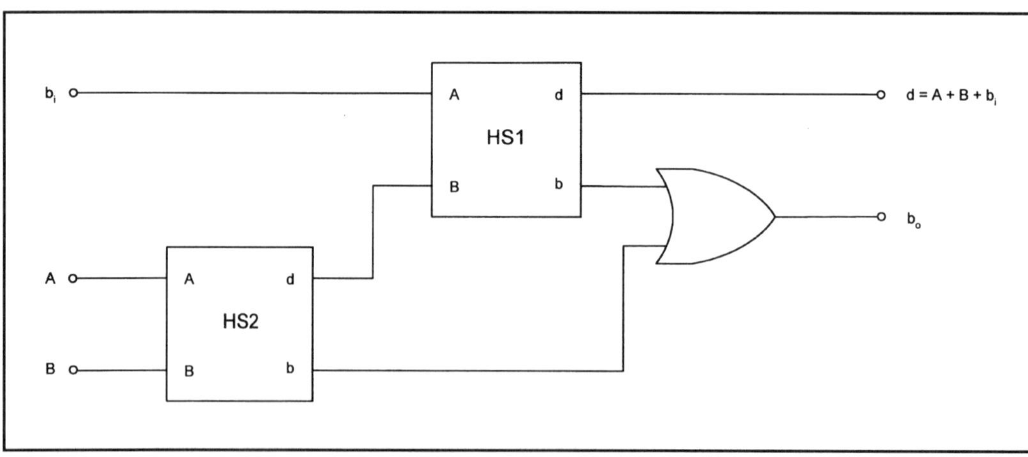

Figure 8-8*. Adder that accounts for carry-in.*

A full-subtractor must not only find the difference between the two inputs and create the borrow operation, but must also account for borrows from other stages. *Figure 8-8* shows the full-subtractor circuit, in which two half-subtractors and an OR gate are used.

A universal adder/subtractor circuit is shown in *Figure 8-9*. Recall that both adders and subtractors produce an output of A XOR B. The principal difference is that adders produce a carry signal (C_o = A AND B) while subtractors produce a borrow signal (b_o = NOT-A AND B).

In *Figure 8-9*, the output of XOR1 will be A XOR B for both addition and subtraction modes. The input of XOR2 (marked "A/S") is the mode control terminal, and selects whether addition or subtraction is used. If A/S is LOW, then the output of XOR2 will be LOW if A = HIGH. The output of the AND gate, then, will be HIGH when A = HIGH and B = HIGH (A X B). This is a carry, so the circuit operates as an adder when A/S = LOW.

When A/S = HIGH, the output of XOR2 will be HIGH when A = LOW. So we can state that the output of the NAND gate is NOT-A X B, which is the definition of a borrow operation. The circuit of *Figure 8-9*, therefore, operates as a subtractor when A/S = HIGH.

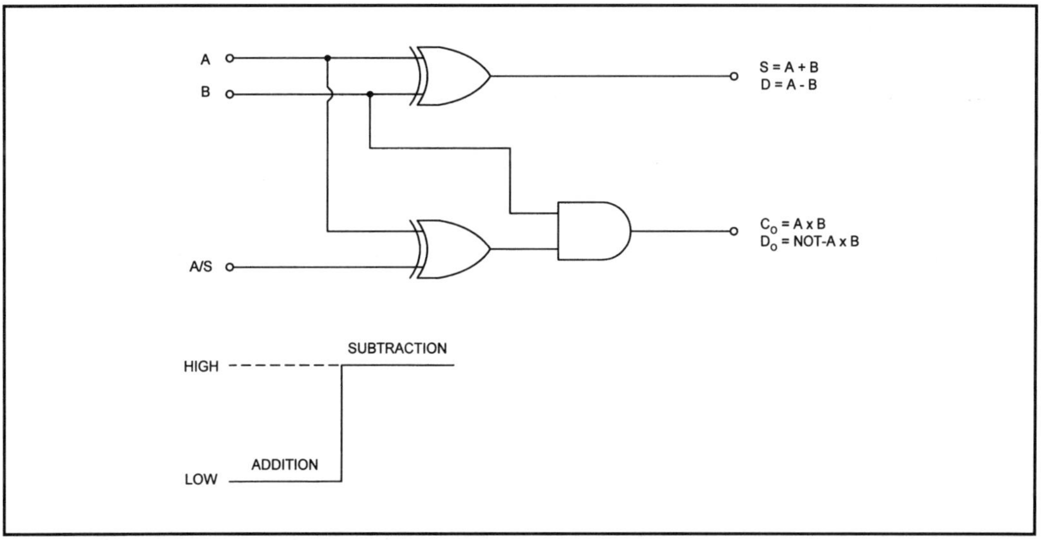

Figure 8-9*. Adder/subtracter.*

Chapter 9

Digital Counter Circuits

A digital counter is a device or circuit that operates as a frequency divider. The most basic form of counter is the J-K flip-flop connected such that the J and K inputs are tied HIGH. This arrangement makes the output produce one output pulse for every two input pulses. It is therefore a binary or divide-by-two counter. We can generalize, saying that a digital counter is a circuit that produces a single output pulse for an x number of input pulses.

You are probably already familiar with digital frequency counters, which are part of test instruments used to measure the frequency of signals. These instruments are decade counters, or divide-by-ten counters. The generic term "counter" applies to a wide range of circuits that fit the definition given above.

There are two basic classes of counter circuits: *serial* and *parallel*. The serial counters are all examples of *ripple counter*. In this type of counter, the output of one stage becomes the input for the next stage. This term means that the input change must "ripple through" all stages of the counter to the proper point. Parallel counters are also called synchronous counters.

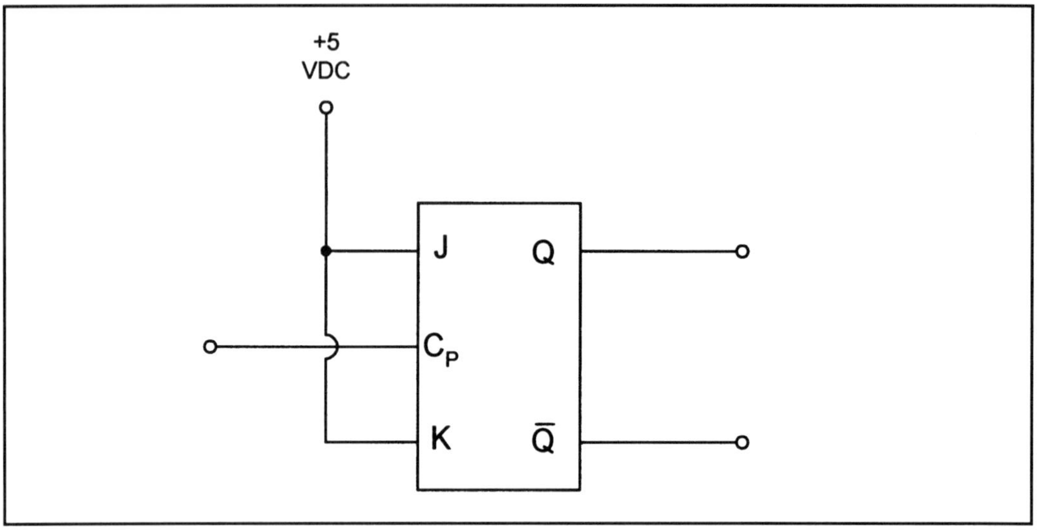

Figure 9-1. J-K flip-flop.

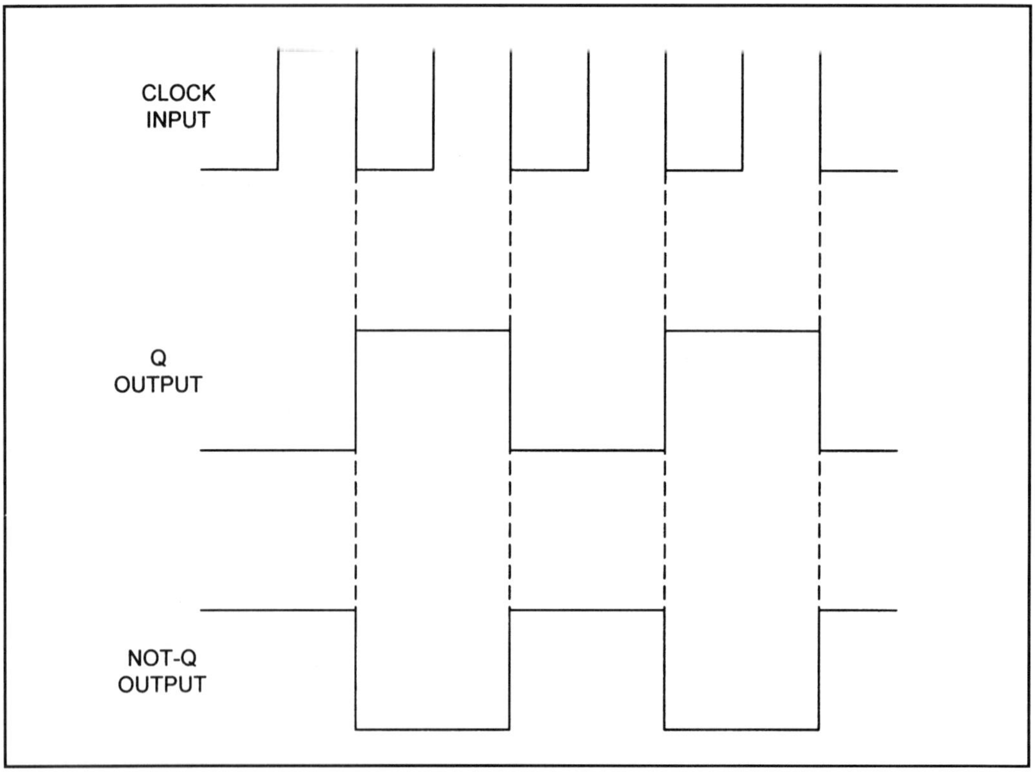

Figure 9-2. Clocked operation of the J-K.

The basic element used in digital counters is the J-K flip-flop, which is shown in *Figure 9-1*. Note in the figure that the J and K inputs are tied HIGH, and therefore remain active.

A timing diagram for the binary counter mode of the J-K flip-flop is shown in *Figure 9-2*. When the J-K flip-flop is connected per *Figure 9-1*, its output changes state on the negative-going transition of the clock pulse. In *Figure 9-2*, the first negative-going clock pulse causes the Q output to go HIGH. It will remain HIGH until the input receives another negative-going clock pulse, at which time the Q output will drop LOW again.

The actions required to complete the output pulse requires two input clock pulses. This J-K flip-flop, therefore, divides the clock frequency by two, making it a *binary* counter.

A binary ripple-carry counter can be made by cascading two or more J-K flip-flops, as shown in *Figure 9-3*. This particular circuit uses four J-K flip-flops in cascade. Any number, however, could be used. A problem with this simple type of counter is that only division ratios that are powers of two (2^N) can be accommodated. In all cases, the division ratio will be 2^N, where N is the number of flip-flops in cascade. In this circuit, therefore, there is a maximum division ratio of 2^4, or 16. Of course, if we can get to the Q outputs of all four J-K flip-flops, we can create frequency division ratios of 2, 4, 8 or 16.

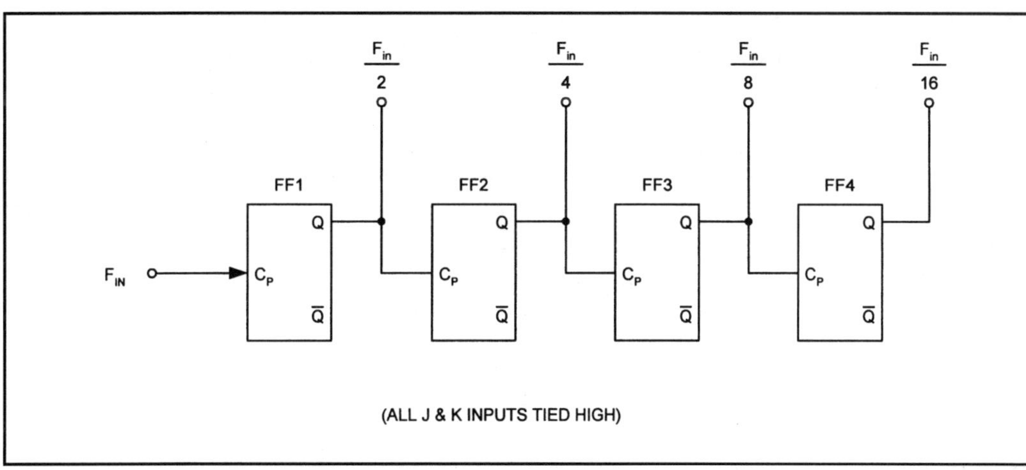

***Figure 9-3**. Four-bit binary counter.*

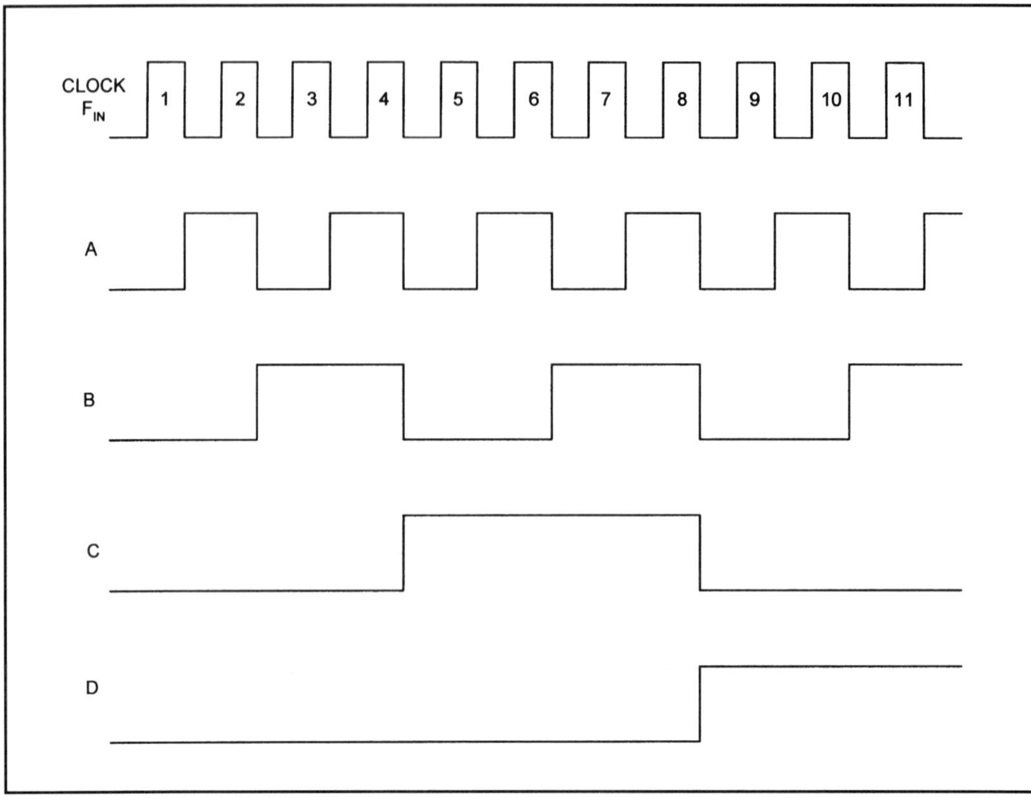

***Figure 9-4**. Counter timing diagram.*

Frequency division is one use for such a counter. You may, for example, want to build a prescaler or some other type of divider. You can use a circuit such as shown in *Figure 9-3* for any division ratio that is a power of two (2, 4, 8, 16, 32 ... 2^N).

But prescaling and most other simple frequency division jobs are but one example of counter applications. We can also use the circuit to store the total number of input pulses; the job most people mean when they think of "counters." Consider again the circuit of *Figure 9-3* and the timing diagram of *Figure 9-4*. Outputs A, B, C and D are coded in binary, with A being the least significant bit (LSB) and D being the most significant bit (MSB). These outputs are weighted in the following manner: $A = 2^0$, $B = 2^1$, $C = 2^2$ and $D = 2^3$. This is called the 1-2-4-8 system because the terms evaluate to A = 1, B = 2, C = 4 and D = 8 in decimal notation.

Recall from our discussions earlier that these are the weights assigned to the binary numbering system (base 2). By arranging the digits in the form DCBA, a binary counter is created that denotes the number of pulses that have been input.

After Pulse No.	Binary Word
0	0000
1	0001
2	0010
3	0011
4	0100
5	0101
6	0110
7	0111
8	1000
9	1001
10	1010
11	1011
12	1100
13	1101
14	1110
15	1111

***Table 9-1**. Binary code conversion.*

Consider the timing diagram of *Figure 9-4*. Note that all of the Q output changes occur following the arrival of each pulse. After pulse number 1 has passed, the Q_A output is HIGH and all the others are LOW. This means that the binary word on the output lines is 0001_2 (1_{10}); one pulse has passed.

Following pulse number 2, we would expect 0010_2 (2_{10}) because two pulses have passed. Note that QB is now HIGH and that all others are LOW. The digital word is indeed 0010_2. If you follow each pulse, then you will find the binary code to be as shown in *Table 9-1*.

The counter shown in *Figure 9-3* could be called a modular-16 counter, a base-16 counter, or a hexadecimal counter. All of these terms mean essentially the same thing.

The output of a hexadecimal counter can be decoded to drive a display device that indicates the hexadecimal digits 0, 1, 2, 3, 4, 5, 6, 7, 8, 9, A, B, C, D and F. In most applications, someone must read the output visually; here, however, a decimal counter is used. We human beings use the decimal number system (probably because of our ten fingers or "digits"). The most significant bit of decimal counters produces one output pulse for every ten input pulses. Decimal counters are also sometimes called *decade* counters. The decimal counter forms the basis for digital event, period and frequency counter instruments.

The hexadecimal counter shown in *Figure 9-3* is not suitable to decimal counting modules unless it is modified by adding a single two-input NAND gate. Recall that a TTL J-K flip-flop uses inverted inputs for the *Set* and *Clear* functions. As long as the *Clear* input remains HIGH, the flip-flop will function normally. When the *Clear* input is momentarily brought LOW, the Q output of the flip-flop goes LOW.

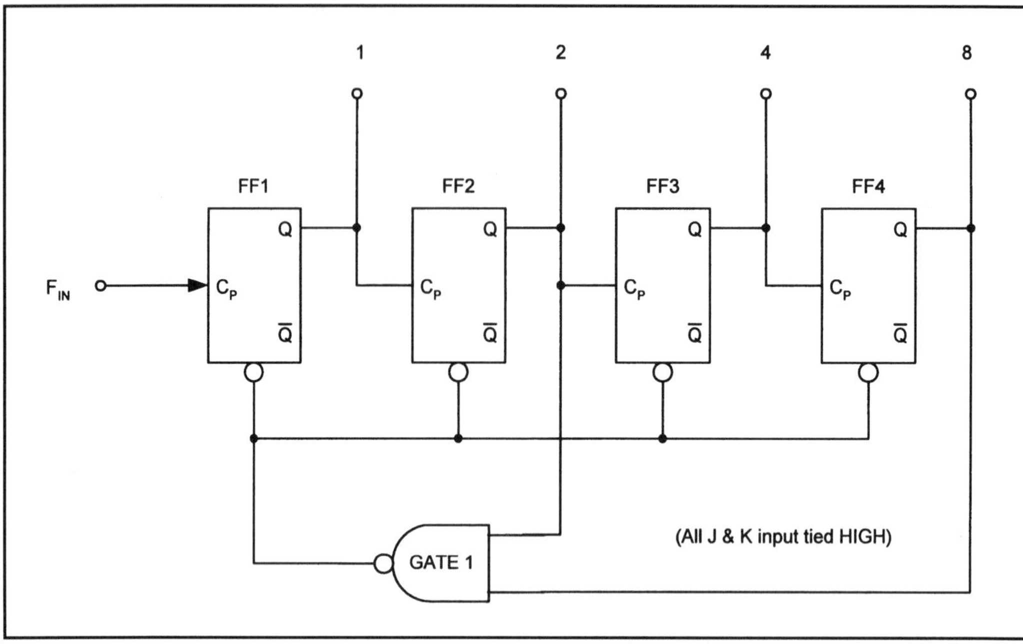

***Figure 9-5**. BCD counter (decimal).*

The decade counter shown in *Figure 9-5* is connected so that all four *Clear* inputs are tied together to form a common *Clear* line. This line is connected to the output of a TTL NAND gate. If either input is LOW, then the output goes HIGH; but if both inputs are HIGH, then the output goes LOW.

The idea behind the circuit of *Figure 9-5* is to clear the counter to 0000 following the tenth input pulse. Let's examine the timing diagram of *Figure 9-6* to see if the circuit performs the correct action. Up until the tenth pulse, this diagram is the same as that for the base 16 counter discussed earlier.

The output of the NAND gate will keep the *Clear* line HIGH for all counts through 10. The inputs of this gate are connected to the B and D lines. The D line stays LOW, which forces *Clear* to remain HIGH, up until the eighth input pulse has passed. At this time (T0 in *Figure 9-6*), D will go HIGH, but B drops LOW. We still have at least one input of the NAND gate (line B) LOW, so the *Clear* line remains HIGH.

The *Clear* input remains HIGH until the end of the tenth input pulse. At that point (T2), both B and D are HIGH, so the NAND gate output goes LOW, clearing all four

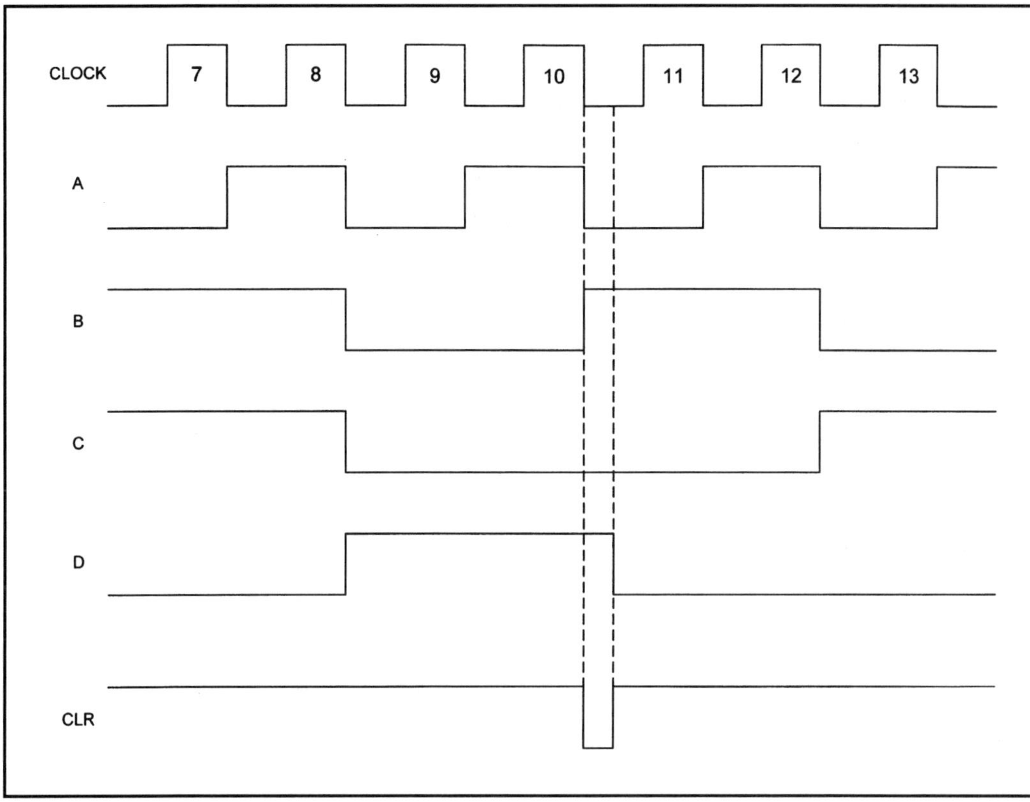

Figure 9-6. *Timing diagram.*

flip-flops. This forces them to go to the state where all Q outputs are LOW. The counter is now reset to 0000.

The reset counter produces a 0000 output, so both B and D are now LOW, and this forces the *Clear* line HIGH once again. The entire reset cycle occurs in the period T2-T3. This period has been expanded greatly in *Figure 9-6* for simplification. Actually, the reset cycle takes place in nanoseconds or, at the most, in microseconds, whatever time is required for the slowest flip-flop to respond to the *Clear* command.

The eleventh pulse will increment the counter one time, so the output will indicate 1 (or 0001_2). This counter, then, will count in the sequence 0-1-2-3-4-5-6-7-8-9-0-1-2 ... The output code is a ten-digit version of four-bit binary, so it is called *binary coded decimal*, or BCD for short.

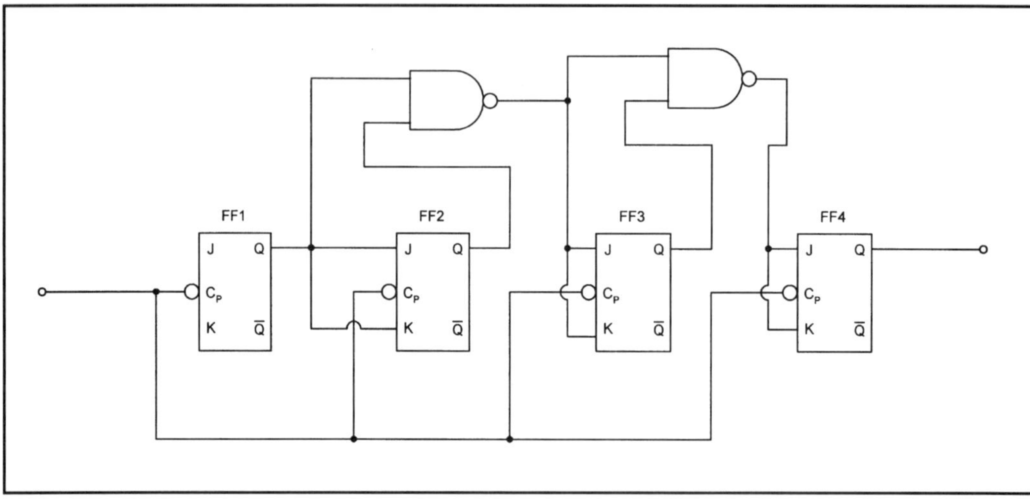

***Figure 9-7**. Counter circuit.*

Synchronous Counters

Ripple counters suffer from one major defect: lack of speed. The counter elements are wired in cascade, so an input pulse must ripple through the entire chain before it affects the output. A synchronous counter feeds the clock inputs of all flip-flops in parallel. The result is a much faster circuit.

Figure 9-7 shows the partial circuit for a synchronous binary counter. Synchronous operation is accomplished by using four flip-flops, with the clock inputs tied together, and a pair of AND gates.

One AND gate is connected so that both Q1 and Q2 are HIGH before FF3 is active. Similarly, Q2 and Q3 must be HIGH before FF4 is made active. On a clock pulse, any of the four flip-flops scheduled to change will do so simultaneously. Synchronous counters attain faster speeds, although ripple counters seem to predominate the common counter applications.

Preset Counters

A preset counter increments from a preset point other than 0000. For example, suppose we want to count from 5_{10} (0101_2). We could preset the counter to 0101_2 and increment from there. The counter output pattern would be:

5 0101
6 0110
7 0111
8 1000
9 1001
0 0000
1 0001
2 0010
3 0011
4 0100
5 0101
—
—
—

Figure 9-8 shows a common method for achieving preset conditions: the jam input. Only two stages are shown here for the sake of simplicity, but adding two additional stages will make it a four-bit counter. Of course, any number of stages may be cascaded to form an N-bit preset counter.

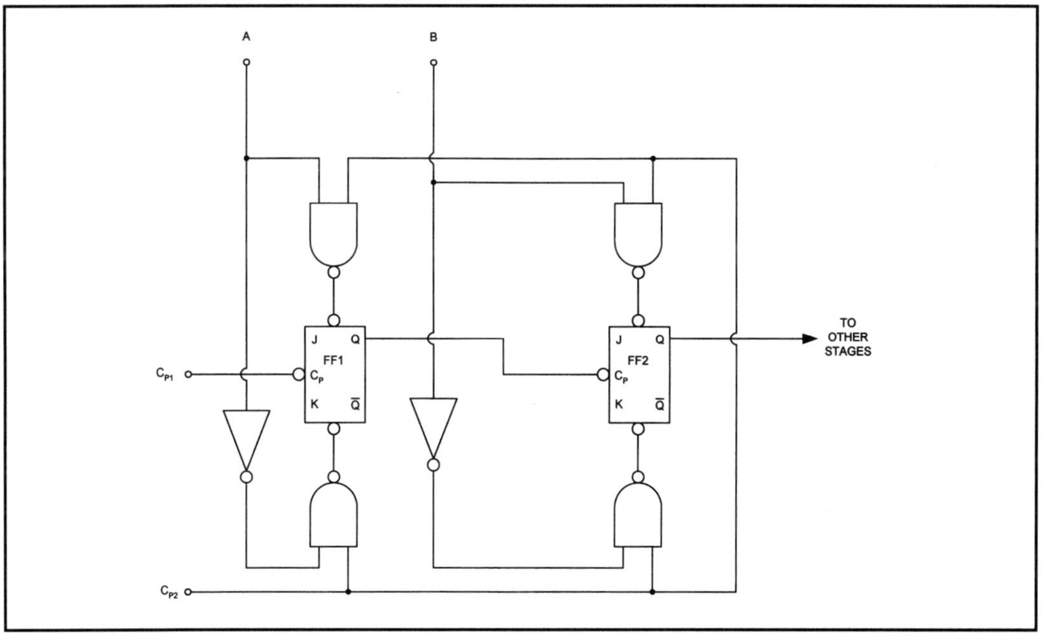

***Figure 9-8**. Preset counter.*

In *Figure 9-8*, the preset count is applied to A and B, and both bits will be entered simultaneously when clock line CP2 is brought HIGH. Line CP2 is sometimes called the *Enter* or *Jam* input. Once the preset bit pattern is entered, the counter will increment from these points with transitions of clock line CP1.

Down Counters and Up-Down Counters

A down counter decrements (instead of incrementing) the count for each excursion of the input pulse. If the reset condition is 0000, then the next count would be $0000_2 - 1_2 = 1111_2$. It would have been 0001_2 in an ordinary up counter. The count sequence for a four-bit down counter is:

0 0000
15 1111
14 1110
13 1101
12 1100
11 1011
10 1010
9 1001
8 1000
7 0111
6 0110
5 0101
4 0100
3 0011
2 0010
1 0001
0 0000
—
—
—

Basically the same circuit is used as before, but toggle each flip-flop from the NOT-Q of the preceding flip-flop. An example of a four-bit binary counter is shown in *Figure 9-9*. Note that the outputs are taken from the Q outputs of the flip-flops, but toggling is from the NOT-Q outputs.

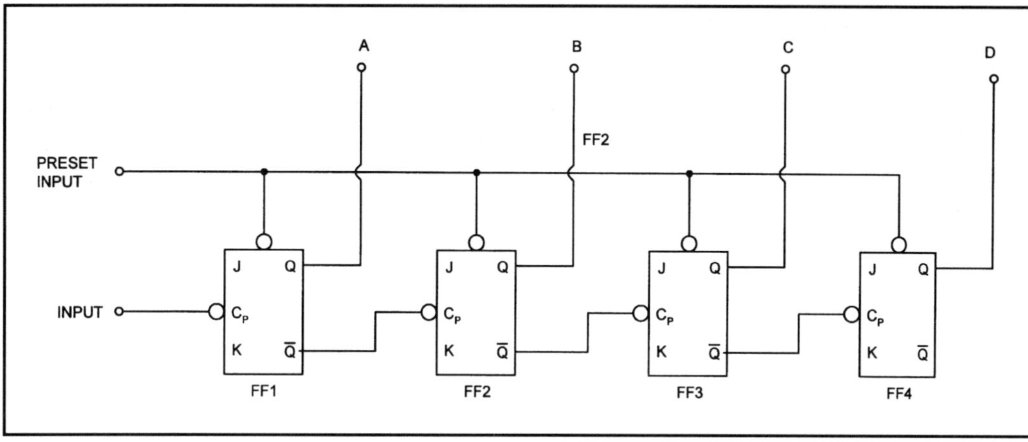

Figure 9-9. *Preset counter.*

The preset inputs of the flip-flops are tied together to provide a means to preset the counter to its initial (1111) state. This counter is also called a *subtraction counter* because each input pulse causes the output to decrement by one bit.

As in the case of the regular decade counter (the up counter), a NAND gate is added to the circuit to reset the counter following the tenth count. The states are detected when C and D are HIGH, and the two middle flip-flops are cleared. This action forces the output to 1001_2 (or 9_{10}). The counter then decrements from 1001 in the sequence:

9 1001
8 1000
7 0111
6 0110
5 0101
4 0100
3 0011
2 0010
1 0001
0 0000
9 1001
—
—
—

Up-Down Counters

Some counters will operate in both up and down modes, depending on the logic level applied to the mode input. If the mode input is HIGH, then the circuit is an up counter. If the mode input is LOW, however, then the circuit is a down counter.

Chapter 10

Display Devices

When digital circuits need to communicate with real, live human beings, some form of display device is required. In this chapter we will take a short look at several different types of display devices. This discussion will not include video monitors, such as those used on computers, but rather electronic components will display alphanumeric or numeric data.

Seven-Segment Displays

The seven-segment display (*Figure 10-1*) is based on the fact that all ten digits (0-9) of the decimal number system can be displayed using just seven bars arrayed in a "figure-8" pattern. The lowercase letters in each of the segments of *Figure 10-1* represent the industry standard designation scheme. This becomes important when interfacing to display driver and decoder circuits.

The ten digits that can be formed using the seven-segment display are shown in *Figure 10-2*. Note that the numbers formed are a bit blocky and are not centered in each cell. The system is, however, able to communicate all ten digits from zero to nine. It falls down markedly, however, when trying to display digits of the hexadecimal system.

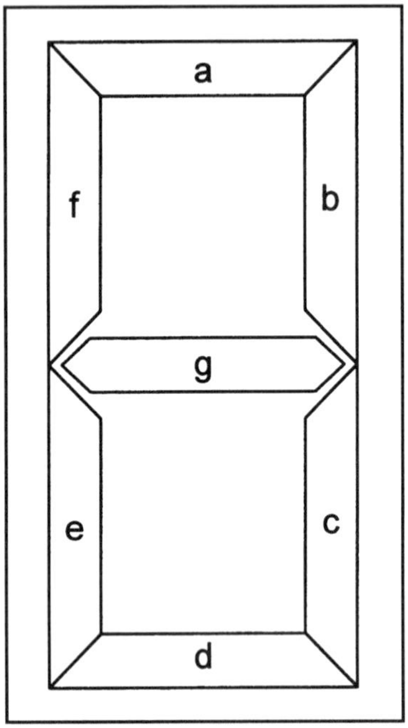

***Figure 10-1**. Seven-segment numerical display.*

The hex system (base 16) uses letters to represent the digits above nine: A = 10, B = 11, C = 12, D = 13, E = 14 and F = 15. The "B" and "D" are difficult to correctly form in seven-segment readouts. The "B" looks exactly like an "8," while the "D" looks like a "0." To display these digits, the lowercase versions ("b" and "d") are used instead of the upper case. This makes it a little difficult to read, but it's workable. What is not workable, however, is the other letters of the alphabet.

Dot Matrix Displays

A method used for making better formed letters is the dot matrix system of *Figure 10-3*. A matrix is formed using rows and columns. The original dot matrix was 5 x 7 (five columns of seven each), but today a number of larger-sized matrixes are used. The matrix shown in *Figure 10-3* is 6 x 9.

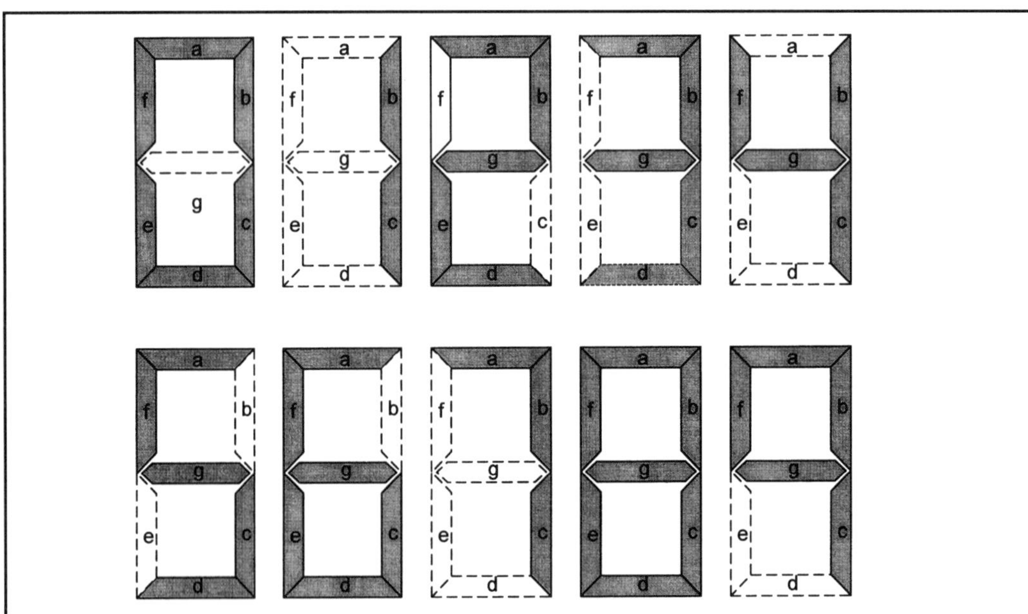

***Figure 10-2**. How the seven-segment display shows numerals.*

The dot matrix was once used extensively in computer printers. A vertical column of seven or more pins was used to strike the ribbon, creating a spot on the paper. When the print head bearing the pins was scanned left to right, and the correct pins punched into the paper, letters, numbers and graphics could be created. The dot matrix printer is now almost universally replaced by laser, ink jet and other printer forms.

The dot matrix concept is also used in computer video displays. If you get a big enough microscope and look at the letters on the screen of a CRT, you will see the same concept in use.

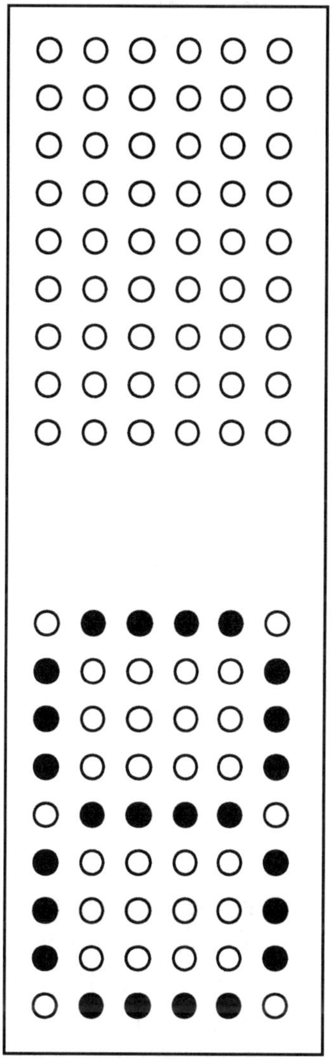

***Figure 10-3**. Dot matrix display.*

Display Device Technologies

A number of different technologies have been used for making displays: neon gas glow lamps, plasma, fluorescent, light-emitting diode and liquid crystal. In this chapter, we will concentrate on the light-emitting diode (LED) and liquid crystal display (LCD) methods.

Light-Emitting Diodes. A light-emitting diode (LED) is a semiconductor component that will emit light when forward biased. The operation of these diodes depends on the fact that atoms raised to a higher energy state by the applied electrical field cannot remain in the higher state for long because of instability. When the individual atoms fall back to their rest or "ground" state, the energy imparted by the electrical field must be given off somehow. In the LED, the energy surrendered when the atoms fall back to the ground state is given off as light.

Recall from above that the seven-segment readout requires seven light sources. When LEDs are used for the light source, then a circuit such as those of *Figure 10-4* are used. There are two versions: *common anode* (*Figure 10-4a*) and *common cathode* (*Figure 10-4b*).

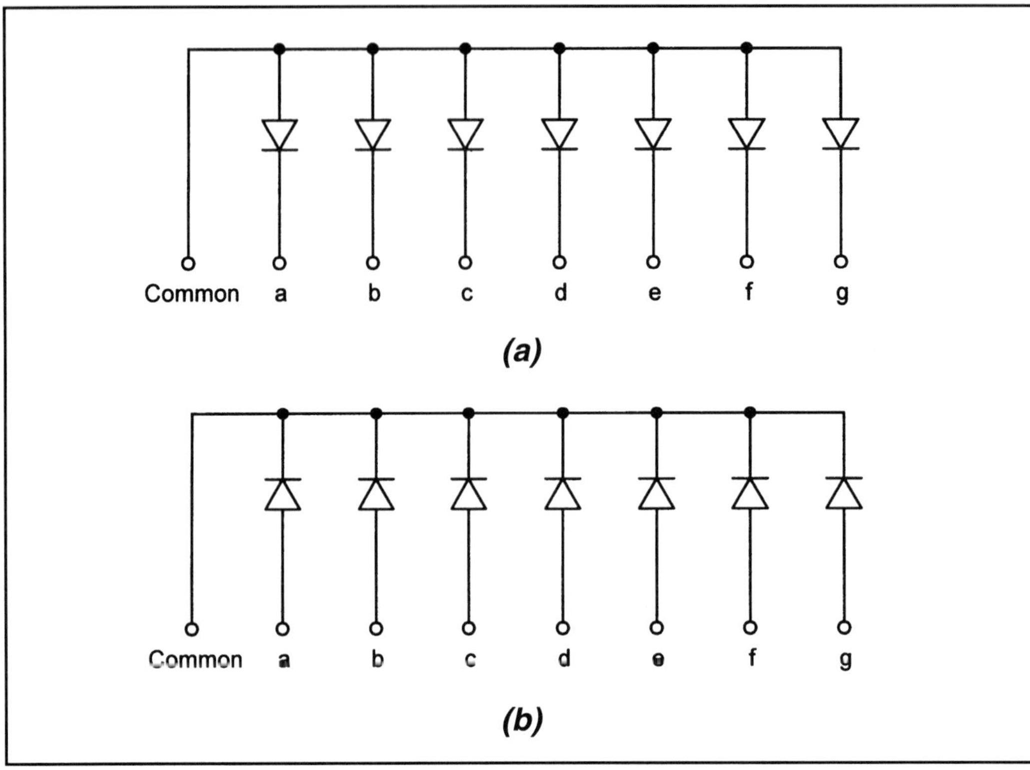

***Figure 10-4**. a. Common anode LED display; b. common cathode.*

Circuit scenarios are suggested by the nature of the common connection. In the common anode form (*Figure 10-4a*), a positive DC potential is applied to the common terminal, then each segment required for the numeral to be displayed is grounded. In the common cathode version (*Figure 10-4b*) the common can be grounded and a positive potential applied to the segment diodes as needed. A series resistor is required for each diode to prevent excessive current from flowing in each diode.

The LED seven-segment display device is available in several forms. Single-digit devices (*Figure 10-5*) are usually equipped with pins that permit them to be either soldered onto a printed circuit board or inserted into a socket. A number of individual displays can be arranged side-by-side (*Figure 10-6a*) to form multi-digit displays. It is also possible to buy multi-digit displays that have all of the LED segments in one component (*Figure 10-6b*).

Liquid Crystal Displays. The liquid crystal display (LCD) depends on the fact that certain chemical compounds have a crystalline nature when an electrical field is

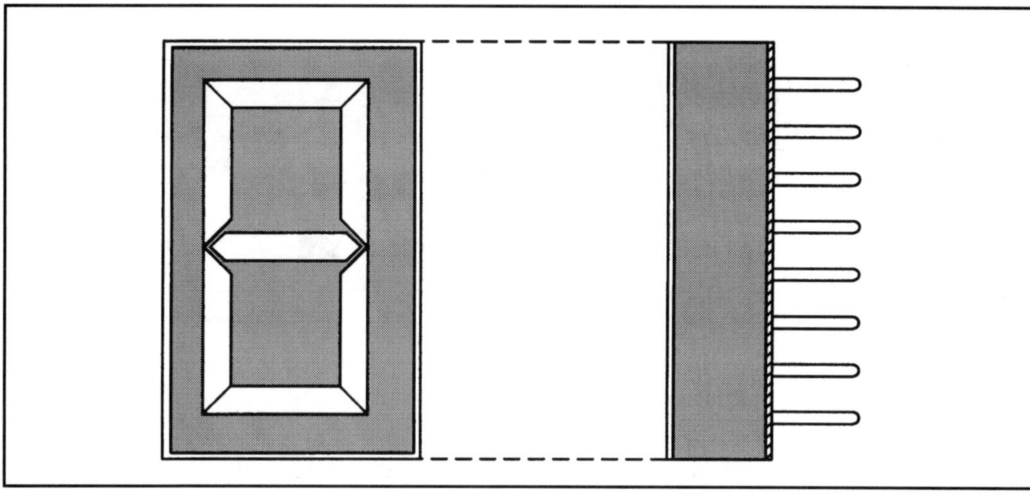

***Figure 10-5**. Seven-segment display device.*

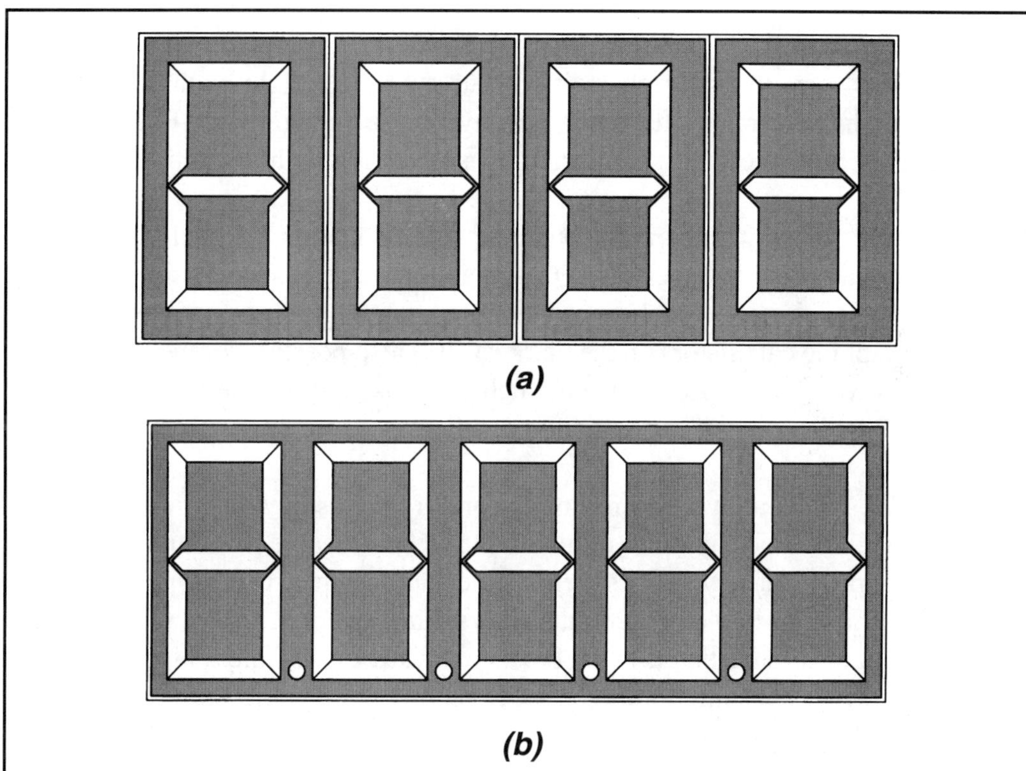

***Figure 10-6**. An array of display devices: a. made of individual displays; b. integrated version.*

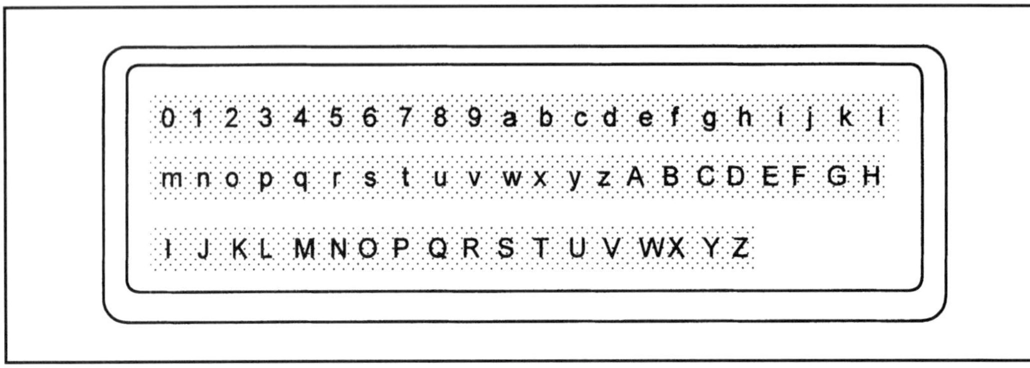

Figure 10-7. Compound LCD display.

applied, and are amorphous when no field is present. By applying the field, the molecules line up to pass polarized light. The LCD digit segment looks light gray until the electrical field is applied, in which case the segment goes dark gray or nearly black.

The LCD devices were once rather difficult to see in low light or direct sunlight. Technological improvements, however, now overcome that problem. Back lighting, for example, permits the LCD segments to be seen in low light conditions.

A very complex display (*Figure 10-7*) can be formed using LCD technology. The manufacturer can provide either discrete alphanumeric cells such as shown in *Figure 10-7*, or more complex displays capable of showing graphical images (note the number of graphing calculators on the market). It is also possible for the LCD manufacturer to place color or special images such as logos on the screen. It is relatively easy to find 80-character LCD compound displays.

LED vs. LCD. There are various reasons to prefer one display over the other. Neither works well in direct sunlight, although the LCD is usable in sunlight while the red LED disappears entirely. Both seven-segment and dot matrix examples are found in both LED and LCD implementation, so this is rarely a concern. However, there is one significant difference that tilts the balance toward LCDs unless high brightness is needed: current consumption. The LCD requires very little current because it depends on voltage fields, not current, to produce the display.

An LED device typically requires 15 mA per segment, so an "8" would require 7 x 15 mA = 105 mA. A five-digit display for a product such as a digital frequency counter would require more than half an ampere of current when "88888" is displayed. This limits the LED display's usefulness in portable equipment, and puts it at a disadvantage in fixed equipment. After all, the DC power supply is a high-cost item as well as a high-maintenance item.

Chapter 11

Registers

A flip-flop is a circuit that is able to store a single bit of digital data. When two or more flip-flops are organized into some configuration to store multiple bits of data, they constitute a *register* circuit. Most registers are, at one level or another, merely arrays of flip-flops.

SISO and SIPO Registers

Figure 11-1 and *Figure 11-2* represent both SISO (serial in serial out) and SIPO (serial in parallel out) forms of register circuit. The only really significant difference is the parallel output lines used on the SIPO device. These lines are absent on the SISO version. Note that many IC registers are either SISO or SIPO depending on how you use them. The SIPO register consists of a cascade chain of type-D flip-flops, with the clock lines tied together — they share a common clock line.

Recall one of the rules for type-D flip-flops: data can be transferred from the D input to the Q output only when the clock input line is HIGH. This rule can be applied to the situation shown in *Figure 11-2*, where the transmission of a single bit of data from left-to-right through the SISO register is shown.

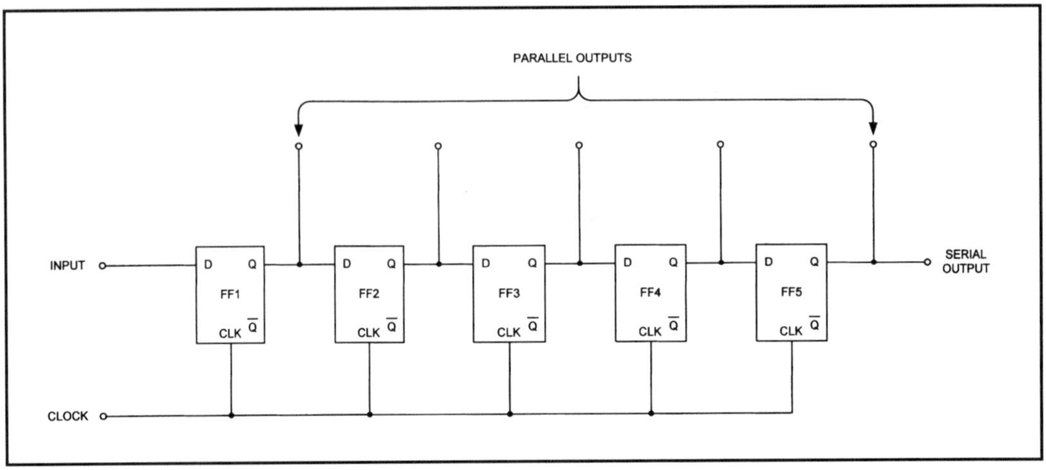

***Figure 11-1**. Shift register circuit.*

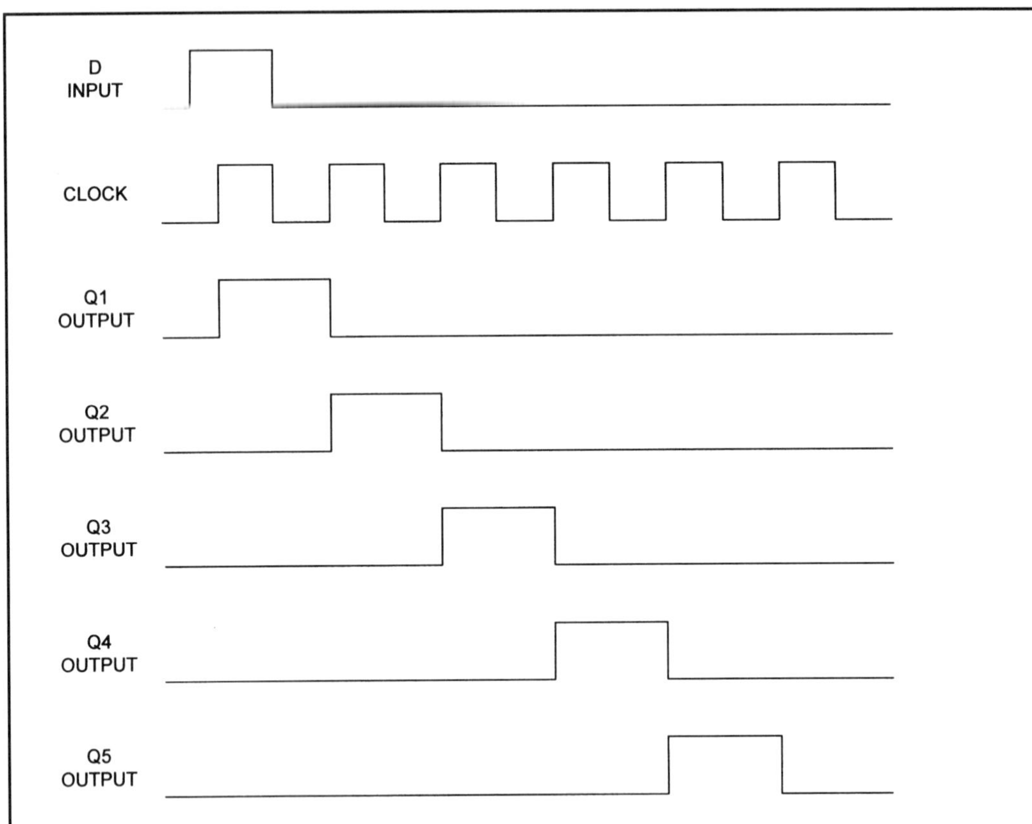

***Figure 11-2**. Timing diagram.*

At the occurrence of the first clock pulse, the input is HIGH. This point is the D input of FF1, so a HIGH is transferred to the Q1 output. This HIGH, which is also applied to the D input of the second flip-flop (FF2), remains after the clock pulse vanishes.

When the second clock pulse arrives, FF2 sees a HIGH on its D input, and FF1 sees a LOW on its D input. This situation causes a LOW at Q1 and a HIGH at Q2.

The third clock pulse sees a LOW on the D inputs of both FF1 and FF2, and a HIGH at the D input of FF3. The third clock pulse, then, causes Q1 and Q2 to be LOW and Q3 to be HIGH.

Note that the SISO input remains LOW after the initial HIGH during clock pulse number 1. This means that the single HIGH will be propagated through the entire SISO shift register, one stage at a time. The HIGH bit will shift one flip-flop to the right each time a clock pulse arrives. If the data at the input had changed, then the bit pattern at the input will be propagated through the shift register.

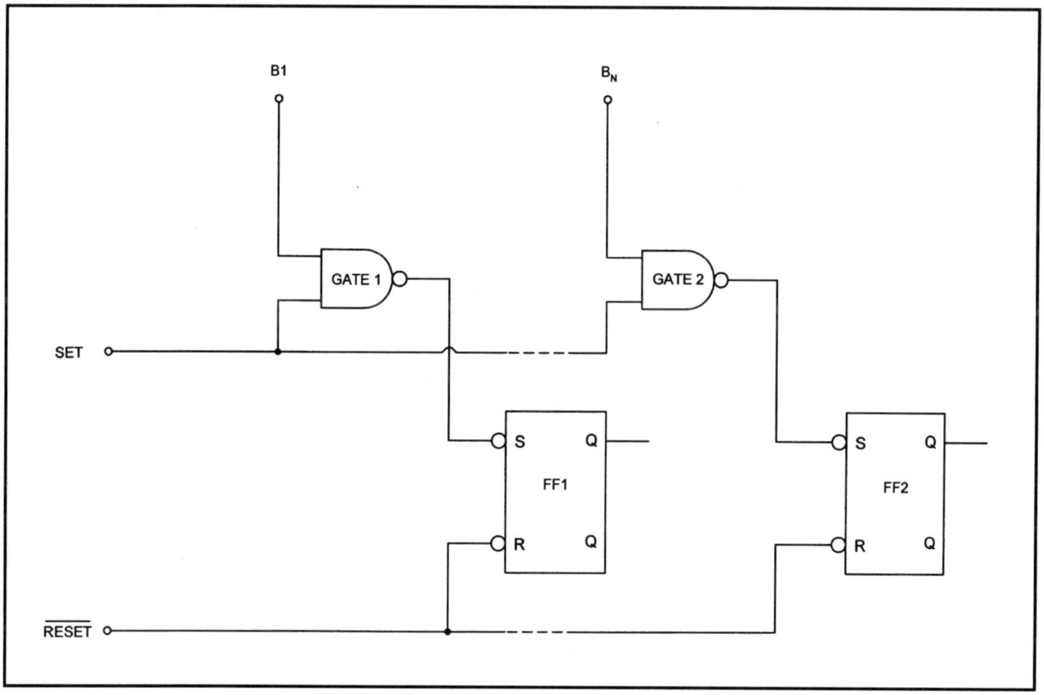

***Figure 11-3**. Preset shift register.*

The shift register shown in *Figure 11-1* is a five-bit (or *five-stage* as it is sometimes called) shift register (although any bit length can be selected). On the sixth clock pulse, therefore, the HIGH is propagated out of the shift register, so all flip-flops are LOW — Q1 = Q2 = Q3 = Q4 = LOW.

The SISO shift register can be made into a SIPO device by adding parallel output lines to the Q outputs of the various flip-flops, or one each for Q1, Q2, Q3, Q4 and Q5. One use for the SIPO shift register is serial-to-parallel code conversion. For economic reasons, digital data is usually transmitted from device to device as a serial stream of bits, especially if long distances are involved. In other words, the bits of the digital word are sent over a communications channel one at a time. But most

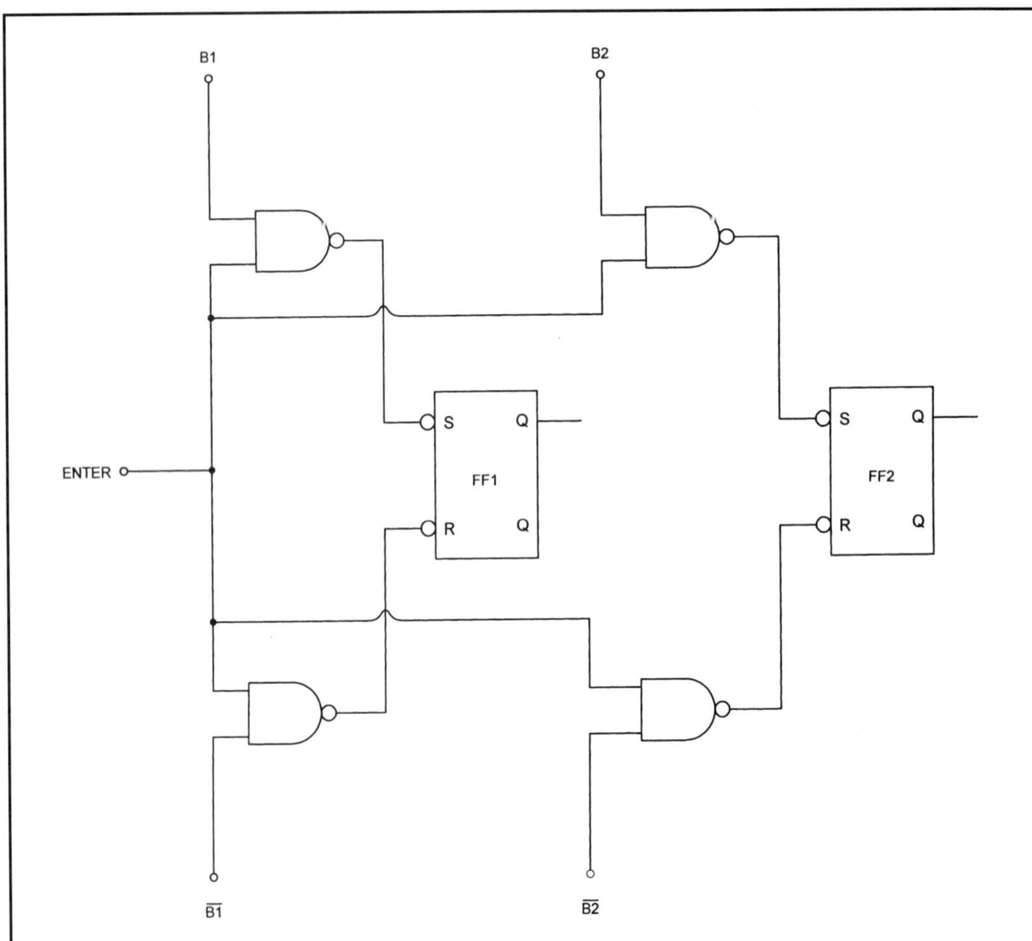

Figure 11-4. *Preset shift register.*

computers and other digital instruments use data in parallel form, which is more important. This format is both faster and more expensive; hence the use of serial data transmission. As an example, consider an eight-bit binary data system. We would need an eight-stage SIPO to convert the eight-bit serial binary code to eight-bit parallel form. The code is entered into the SIPO register one bit at a time, so that after eight clock pulses, the first bit will appear at Q8, and the last bit at Q1.

Parallel Entry Shift Registers

Parallel entry shift registers are faster to load than serial input types because a given bit in the shift register can be changed directly, without going through all the other stages. In other words, the new bit need not ripple through all the stages to reach its destination stage. There are two basic forms of parallel entry shift register: *parallel direct* and *parallel jam*.

In parallel direct entry, or simply parallel entry (shown in the partial schematic of *Figure 11-3*), the register must first be cleared by setting all bits to LOW and first bringing the Reset line momentarily LOW. The data that is applied through inputs B1 through B_N can be loaded into the register by momentarily bringing the *Set* line HIGH.

The jam entry circuit of *Figure 11-4* is able to load data from the B1 through B_N inputs in a single operation. This job is accomplished by loading the complements of the B1 through B_N bits into the other inputs. In all IC shift registers using this technique, inverter stages are connected internally to the inputs in order to automatically complement the input lines. The outside user never sees the complementing process.

A recirculating shift register circuit is shown in *Figure 11-5*. These registers are able to read out the data in serial format, then re-enter it so that it is retained. This is done by connecting the output of the SISO register back to its input. As the data is read out of an ordinary SISO register, it is destroyed and lost forever. However, we sometimes wish to retain the data and be able to read it out. The solution here is to read it right back in as it is read out. Although only three bits are shown here for simplicity, the bit length could be almost anything.

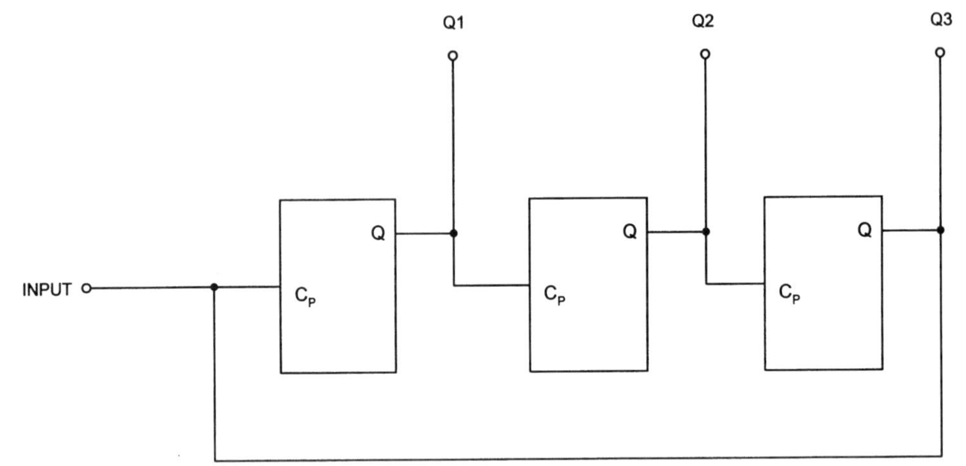

***Figure 11-5**. Recirculating shift register.*

One interesting application for the recirculating shift register is in modern medical "non-fade" and other storage oscilloscopes. Most of the human physiological signals that physicians want displayed on a scope are very slow; that is, a 1 Hz repetition rate represents a heart rate of 60 bpm. The trace tends to fade before the entire waveform complex is displayed. Modern oscilloscopes overcome this problem by digitizing the ECG waveform then storing it in a recirculating shift register. By scanning the memory (which means reading out the serially stored data into a digital-to-analog converter), the entire waveform can be displayed on the screen of the CRT at one time. This is done 64 to 1,000 times per second, depending on the model and the fidelity required. New waveform data is added by replacing the four to ten oldest bits with new data as it occurs.

IC Examples

Few designers would bother using an array of cascade flip-flops in making a shift register. There are simply too many different types of shift registers on the market in both TTL and CMOS families.

Chapter 12

Data Multiplexers and Selectors

The fastest way to transfer multibit data from one part of a digital circuit to another is to connect all of the like-bits in parallel — all B0 lines in parallel, all B1 lines in parallel, etc. But parallel data transfers are somewhat expensive in cost and in printed circuit board "real estate." They can become terribly expensive in hardware costs when there is more than one data source. If, for example, there are three channels of 16 bits each, a total of 48 parallel lines may be required.

This might not seem like much. After all, any technician worth a shaker full of salt can easily and correctly connect 48 wires in very short order. But these wires must be viewed in a different light once outside the same room. Digital signals cannot be passed along very long wires without help. The problem is the sharp rise time of the pulses used in digital electronics. Low impedance line drivers and line receivers must be used at the input and output ends of the line, respectively. The "wires" must be viewed as communications channels instead of as mere lengths of copper wire. When you go outside the immediate vicinity and have to rely on the services of telephone common carriers or Internet service providers, then these are indeed communications channels. How would you like to pay for 48 radio or wire communications channels? The cost is simply too much for this type of data communication.

The solution is in serial data transmission, which is a topic we will examine later in this book. For the present, however, we will cover how to make several data channels fit into one communications channel, or how to make parallel data lines into serial data lines. The answer is in *multiplexing*.

Multiplexing

Multiplexing is the art of interleaving data so that the data channels can be transmitted together over a single line or passed through a single communications channel. Two basic forms of multiplexing are *frequency domain* and *time domain* (there is also a *code domain*, but we will not discuss this one). The most common examples of frequency domain are color television and FM stereo broadcasting. All of these stations use one or more *subcarriers* to transmit information. Digital systems, however, can use time domain "muxing."

Figure 12-1 shows a simplified example of how a time domain multiplexer circuit works. Here, two rotary switches are used to select from any of six different data channels. These switches are rotated *synchronously* so that both are always in the same position; in other words, S1 is connected to channel 3 at the same time that S2 is connected to channel 3. This situation allows us to transmit data from six channels serially along the communications paths by using techniques that will be discussed more fully in the next section.

Figure 12-2 shows a system in which four 16-bit devices are connected to a single 16-bit input port. If they were connected in parallel, then 64 lines would be required

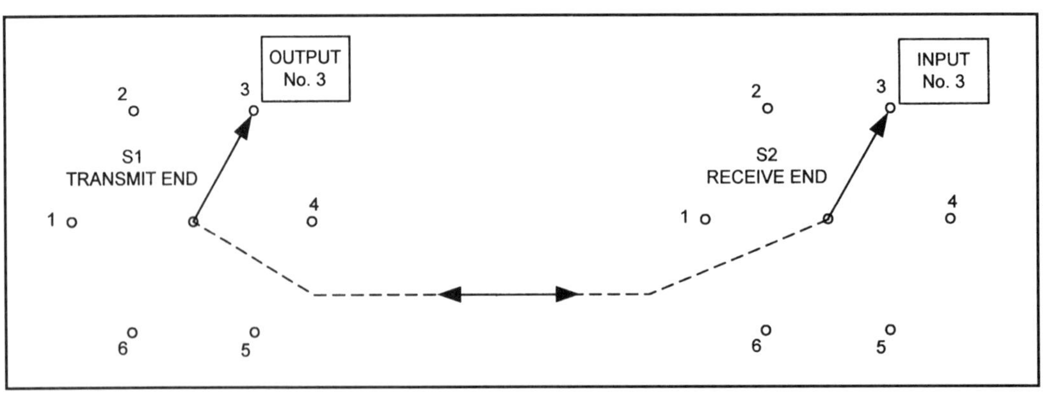

***Figure 12-1**. Equivalent circuit of a multiplexer.*

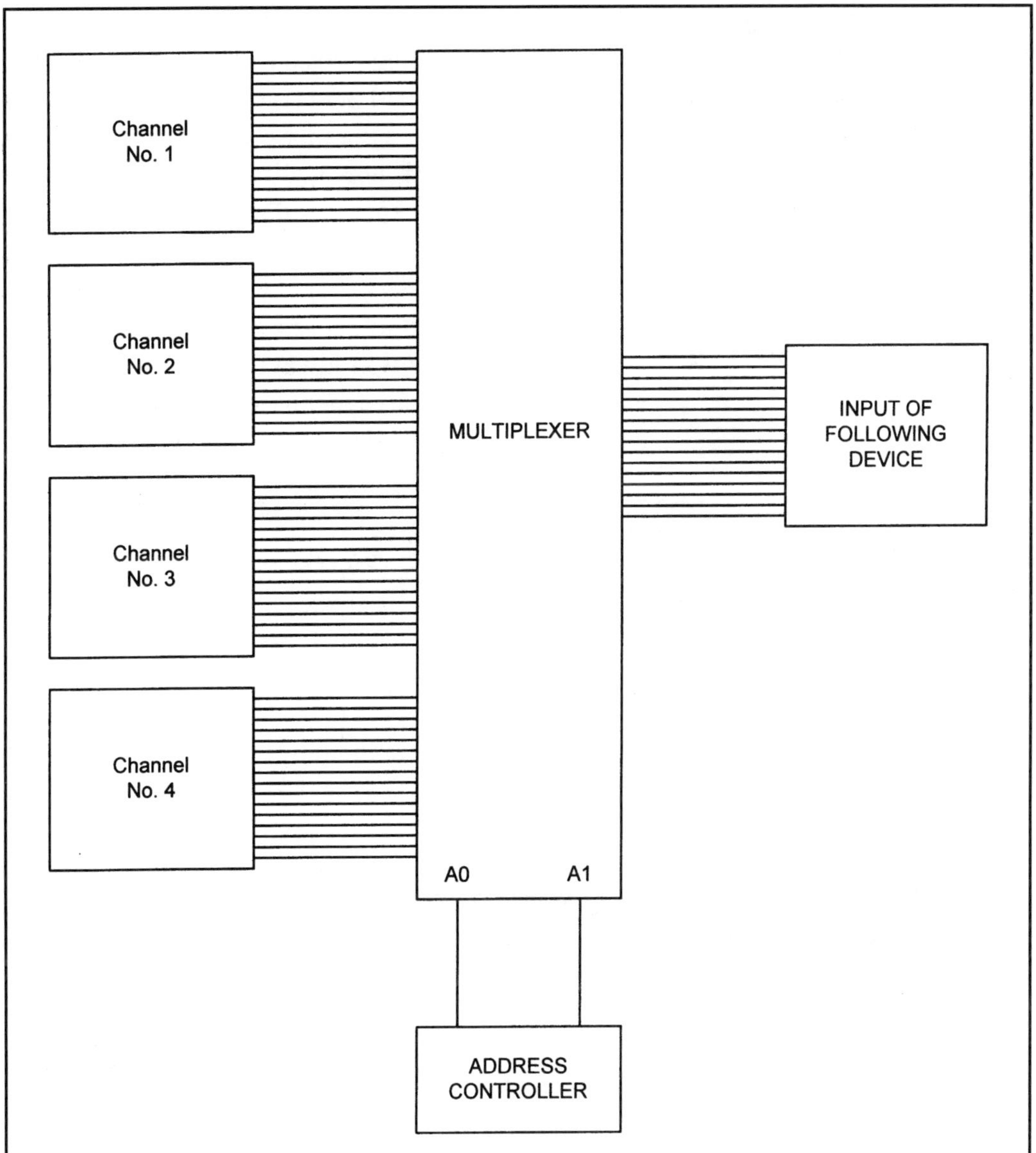

***Figure 12-2**. Block diagram of a multiplexed system.*

instead of 16. The remaining two wires are used for control of the multiplexer. In the circuit shown, the multiplexer can be viewed as a programmable switch. These two wires do the programming according to the rules shown in *Table 12-1*, using binary codes from 00_2 to 11_2.

CHANNEL	CODE
1	00
2	01
3	10
4	11

***Table 12-1**. Rules for programming the multiplexed system, using binary codes from 00_2 to 11_2.*

If the code is 00, then channel 0 is selected and connected to the data bus. If the code is 01, then channel 1 is selected, and so on. A simple two-bit binary counter (or a counter program if a computer or PIC controller is used) will create the addressing codes. It will sequence 00-01-10-11-00 and so forth until turned off.

The circuit shown in *Figure 12-2* assumes that the data at the inputs will always be ready because it operates in a quasi-asynchronous mode. Synchronous operation would require at least one more wire.

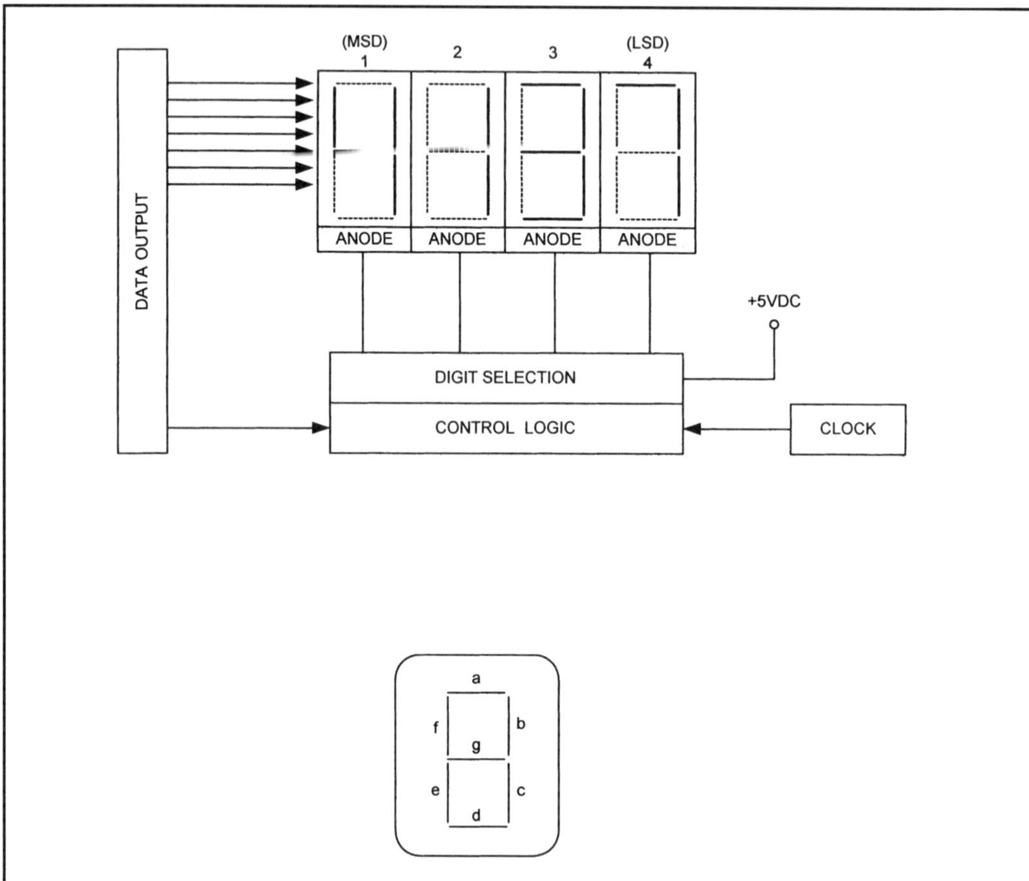

***Figure 12-3**. Seven-segment displays multiplexed.*

Of course, it must also be assumed that the device receiving the data will not change the data selector code until it has successfully completed the current data transfer. It will not, for example, switch to channel 2 until it has successfully input the data on channel 1

Figure 12-3 shows another form of digital multiplexing once used extensively on old calculators and other portable digital instruments. LED and certain other seven-segment display readouts can draw considerable amounts of current from the DC power supply. Some larger devices can draw up to 40 mA per segment. In the four-digit display shown in *Figure 12-3*, when all segments are turned on (displaying the value "8888"), the current drawn is 1.28 amperes.

If this were a 12-digit calculator (or other display), then even more current (4 amperes) would be needed. Once you begin to require so much current on a continuous basis, battery life drops and heat dissipation rises. Both are costly, and in this case, "portability" becomes questionable.

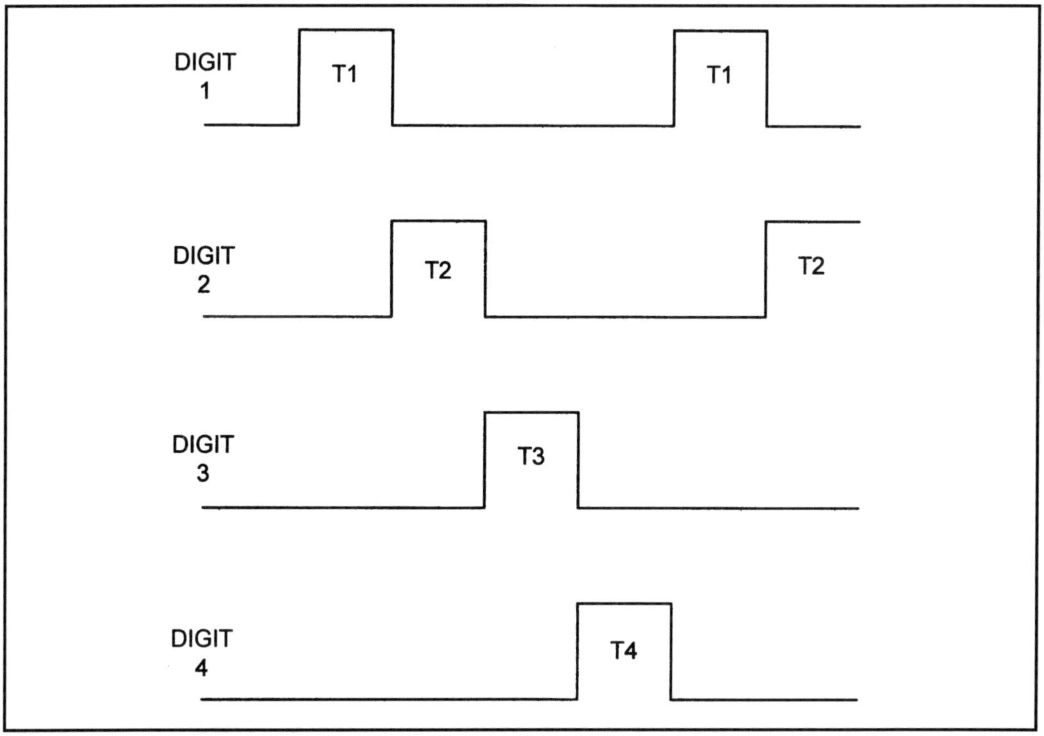

***Figure 12-4**. Timing diagram.*

The answer to the battery life dilemma is to multiplex the digits. Only one digit at a time will be turned on, but if the switching rate is fast enough, then the *persistence* of the human eye will make the display appear to be constant. This "flicker fusion frequency" tends to be around 8 to 10 Hz in most people.

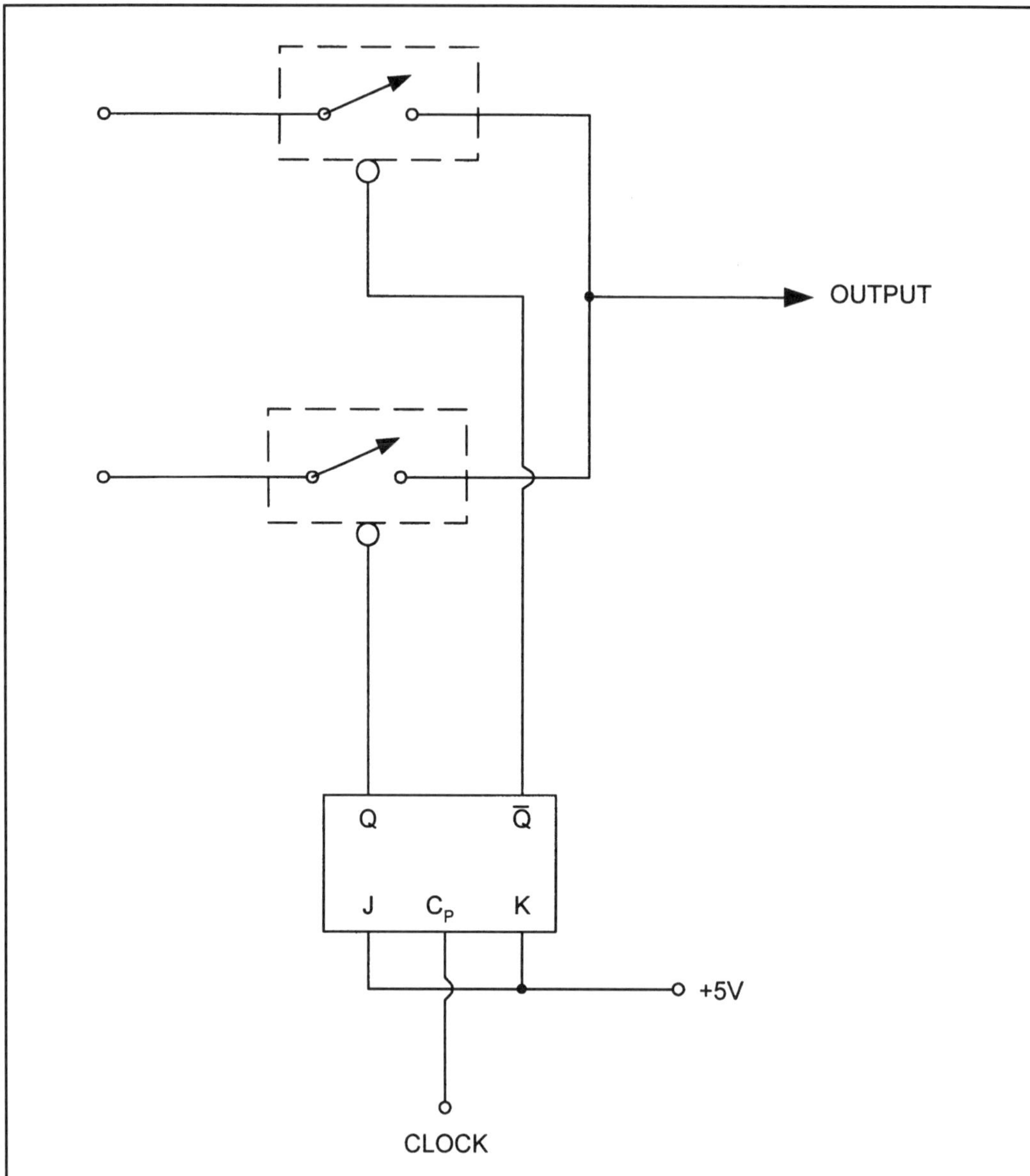

Figure 12-5. *Analog/digital switches.*

Figure 12-3 shows a multiplexed display, while *Figure 12-4* shows the timing diagram. Recall that a seven-segment readout requires one wire for each segment (plus at least one common wire). These are typically labeled a, b, c, d, e, f and g (see inset to *Figure 12-3*). In a typical device, power is applied to the anode of the digit, and the decoder will ground the terminals to those segments that are to be lighted. To display a "3," for example, grounds the terminals a, b, c, d and g. The instrument will place the proper seven-segment code for the digit to be displayed on the seven-bit data bus at the appropriate time. Timing is critical to the proper operation of this system, and it is handled by the control logic section.

The *digit selector* is a circuit that enables each digit in sequence. In the crude example shown here, a digit is selected by applying +5VDC to the common anode. This circuit has four output lines, one for each digit. When a selector line is HIGH, then +5VDC is applied to its digit.

In the situation shown in *Figure 12-3*, the four-digit number "4137" is displayed. The sequential operation is as follows:

Time T1. The seven-segment code that represents "4" is placed on the data bus. Simultaneously, the digit selector applies +5VDC to digit number 1. This causes digit number 1 to display a "4."

Time T2. The seven-segment code that represents "1" is placed on the data bus. Simultaneously, the digit selector applies +5VDC to digit number 2. This causes digit number 2 to display a "4" and turn off all other digits.

Time T3. The seven-segment code that represents "3" is placed on the data bus. Simultaneously, the digit selector applies +5VDC to digit number 2. This causes digit number 3 to display a "3" and turn off all other digits.

Time T4. The seven-segment code that represents "7" is placed on the data bus. Simultaneously, the digit selector applies +5VDC to digit number 4. This causes digit number 4 to display a "7" and turn off all other digits.

This sequence happens continuously. Only one digit is turned on at any one time, but they are all turned on sequentially many times per second.

A simple example of a two-channel multiplexer is shown in *Figure 12-5*. This circuit uses a common CMOS digital integrated circuit, the 4066. Note that the CMOS electronic switches used for this example are bilateral devices; that is, they will pass current in either direction. This means that they can be used as either analog (where bipolar signals are encountered) or digital multiplexers. Each of the switches inside of the 4066 device is completely independent of the others except for the DC power supply and ground terminals.

The control pins are 5, 6, 12 and 13. When the voltage applied to these pins is equal to the voltage on pin number 7, the switch selected by that particular pin is turned off. In digital IC devices, pin number 7 is often used for ground (especially in 14-pin DIP packages). The switches are turned off when their control pins are grounded (or at zero volts). The voltage applied to pin number 7 can be anything in the range 0V to -5V (or -7V in some models), but the negative voltage is seen only occasionally in digital circuits.

Applying a voltage equal to the voltage on pin number 14 to a control pin will turn the switch on. In most digital circuits, this will be +5V, although CMOS devices can operate at voltages up to +18V. The rationale is to make the whole circuit TTL compatible.

The terms "on" and "off" refer to the series impedance of the switch. In the off condition, the series impedance is very high, usually well over one megohm. In the ON condition, on the other hand, the series impedance drops very low, always less than 2,000 ohms and in some devices less than 100 ohms.

In the multiplexer circuit of *Figure 12-5*, only two of the switches are used. This circuit, incidentally, is also useful for making a single trace oscilloscope into a dual trace oscilloscope. Note that one terminal of both S1 and S2 are connected together to form a common output. The remaining terminal on both switches goes to the respective channels. Pin number 1 is connected to channel 1, and pin number 4 is connected to channel 2.

The switching action is performed by a J-K flip-flop connected in the clocked binary divider mode. The Q and NOT-Q outputs on a J-K flip-flop are complementary. This means that one output will be HIGH and the other is LOW at any given point in time. When the state of the flip-flop changes, the HIGH output becomes LOW and

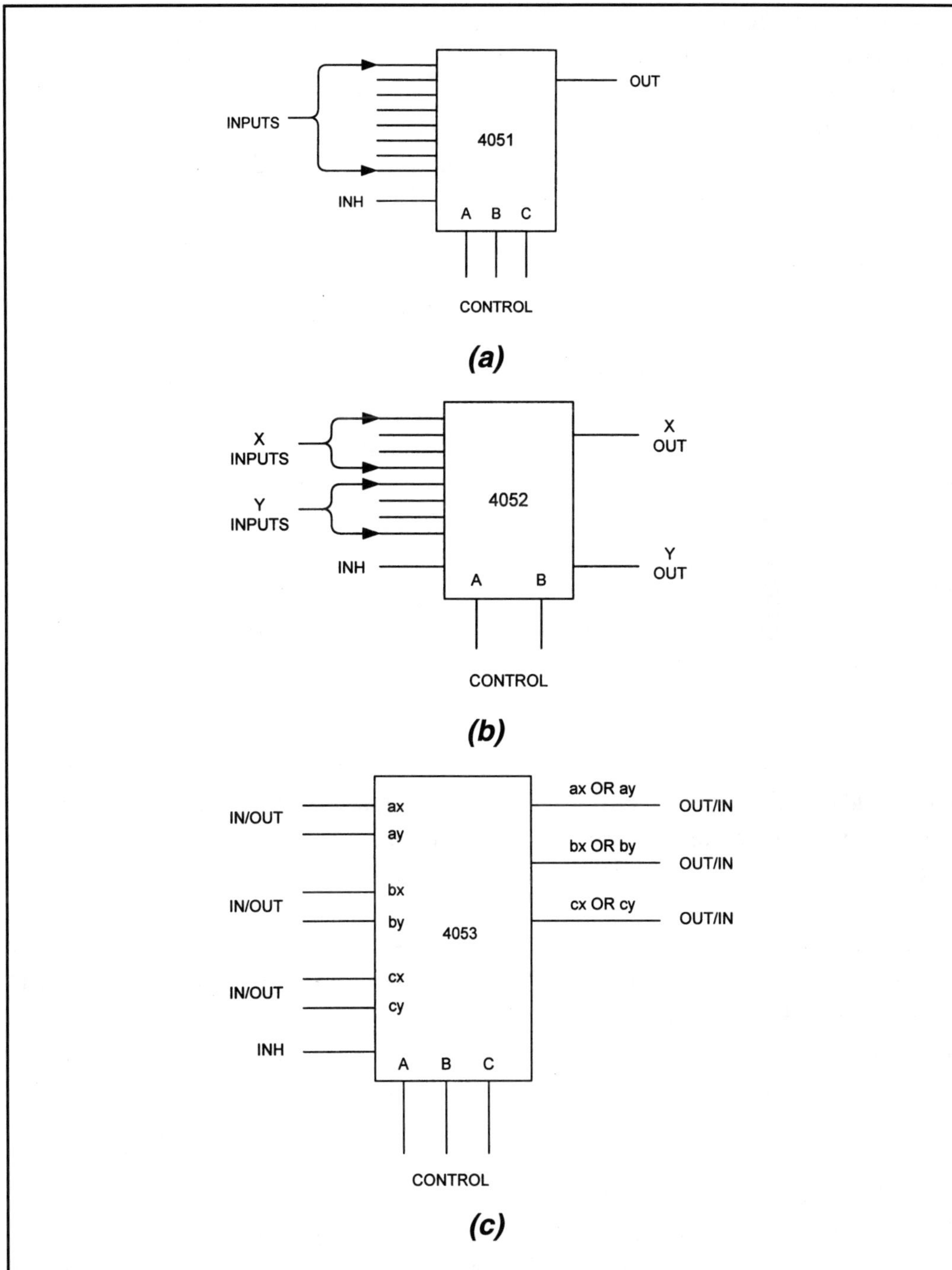

***Figure 12-6**. Examples of analog/digital switch.*

the LOW output becomes HIGH. The timing diagram shown in *Figure 12-5* shows this relationship more clearly.

During interval T1, the Q output is HIGH and NOT-Q is LOW. This turns on switch S1 and turns off S2 (pin number 5 is HIGH and pin number 13 is LOW). According to the rules of operation for the 4066 device, channel 1 is then connected to the output.

The switches are turned on and off alternately. This means that the output will be connected to the respective inputs alternately.

In the case of an oscilloscope switch, only the one circuit is needed to make things happen. In data communications systems, however, the signals are interleaved, which makes things a hopeless timing nightmare at the other end of the transmission path unless some form of decoding is provided. A *demultiplexer* circuit must be provided at the receiver end of the path.

Fortunately, the 4066 devices are bilateral. Basically the same circuit is also useful for demuxing the data stream. The serial data stream coming into the receiver is fed to the common terminal between the two switches (used as an output at the mux end and an input at the demux end).

Three additional multiplexers and demultiplexers are 4051, 4052 and 4053. All three of these CMOS devices have *chip enable* pins that allow the chip to be turned on and off at will. The 4051 shown in *Figure 12-6* is a 1-of-8 selector and can be viewed as analogous to a single-pole eight-position rotary switch. Like the 4066 device, the 4051 is bilateral. Three control terminals (A, B and C) select which one of the switches is turned on at any given time. These are programmed according to the standard rule.

Chapter 13

Monostable Multivibrators

The *monostable multivibrator*, also called the one-shot or MMV, is a digital circuit that has only one stable output state and a trigger input (see *Figure 13-1a*). When a trigger is received, the MMV will snap to its unstable state, remain there a preset period of time, then automatically revert to the stable state. *Figure 13-1b* shows the timing situation where the MMV is stable when the output = LOW, and unstable when the output = HIGH. The output remains in the dormant condition, with output = LOW, until time T1 when a trigger pulse is received by the trigger input. The output snaps HIGH, and remains HIGH for a predetermined period of time, called the *output duration*, T.

In this example, both the output pulse and trigger pulse are positive-going, or their normal rest condition is LOW. However, it might also be that either or both (trigger/output) may be HIGH and drop LOW when activated. One popular combination is to have a LOW-to-HIGH transition on the output, and a HIGH-to-LOW transition on the trigger pulse. This is seen in LM-555 IC timer MMVs and operational amplifier MMVs.

One of the features of the standard MMV is that it will not respond to additional trigger pulses until the output duration T has expired. This action is seen in *Figure 13-1c*. At T1, a trigger pulse is received, and the output snaps HIGH and remains

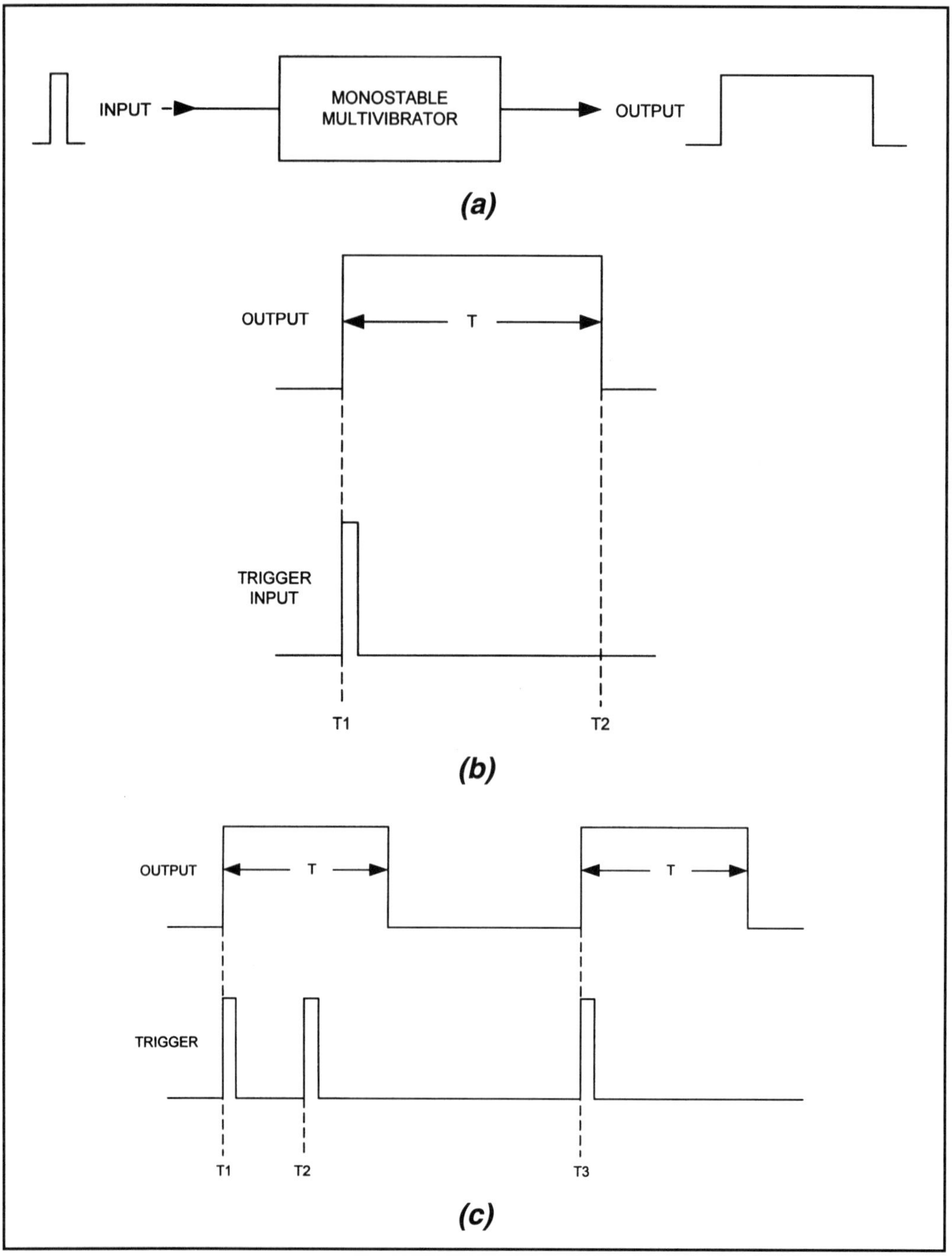

***Figure 13-1**. a. Monostable multivibrator; b. timing diagram; c. timing diagram showing non-retriggering.*

HIGH for period T. At T2, a second trigger pulse is received, but the MMV ignores it. At T3, after the duration, T, of the first output pulse "times out," a third trigger pulse is received. Because the output is back in the dormant stable state, the action will occur all over again. A second output pulse of duration, T, is generated. Later, I will discuss a class of circuits called *retriggerable MMVs*, which will respond to trigger pulses that occur before time-out.

Applications for the Standard MMV

So, what use is a circuit that produces a single output pulse for a single input pulse? This seems like a waste of effort on first glance. But there are actually quite a few applications for the MMV. One common application is keyboard and switch debouncing. The contacts inside mechanical switches, keyboards and electromechanical relays tend to bounce when actuated. As a result, for several milliseconds, a series of on-off conditions exists, producing a noise component on the circuit being switched. You can see this effect quite nicely on an oscilloscope or by connecting a switch, resistor and voltage source to the input of a binary counter. The counter will toggle eight or ten times on the average switch before the action dies out. By using the first "bounce" to trigger a one-shot (which has a duration longer than the bounce time of the switch), the circuit will produce a single output pulse for every operation of the switch.

Another application is pulse stretching. If a pulse produced by a source is too short, then it can be used to trigger an MMV that has a longer pulse duration. The MMV can also be used to do input signal cleanup. If a pulse source produces a ragged-looking pulse, or if the pulse is distorted in transmission (which often happens in

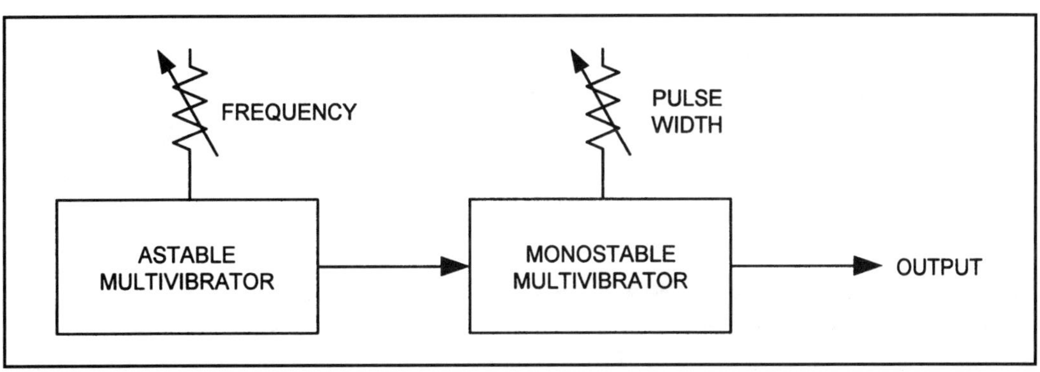

Figure 13-2. *Pulse generator.*

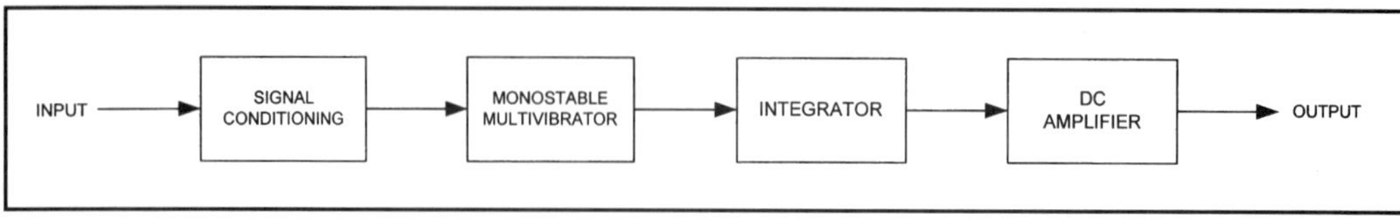

Figure 13-3*. Tachometer circuit.*

coaxial cable systems), then the MMV can be used to create a new pulse on a one-for-one basis. This will, in effect, "rejuvenate" the pre-existing pulse.

Figure 13-2 shows the use of an MMV, along with an oscillator ("astable multivibrator") circuit to create a variable frequency, variable pulse-width pulse generator. The basic frequency is set by the oscillator circuit. The output signal from the oscillator is used to trigger the input of the MMV. By making the timing resistor variable, the MMV will produce a variable-width output pulse with the same frequency as the input frequency

Figure 13-3 shows an analog tachometer based on the MMV circuit. A tachometer produces a DC output voltage that is proportional to the frequency, or repetition rate, of the input signal. Sometimes, the signals at the input are a little ragged, so some form of signal conditioning is needed. A comparator or Schmidt trigger, for example, will produce output pulses from almost any shape of input signal, with a frequency equal to the input signal. But those pulses are not suitable for frequency measurement. The problem is that the duration of each conditioned pulse varies. By using them to trigger an MMV circuit, however, a train of constant duration and amplitude pulses is produced that vary in repetition rate or frequency with the input signal.

The MMV tachometer uses an integrator circuit to convert the input pulses to a DC voltage proportional to the input frequency. The integrator output is a function of the input pulse durations, the input pulse amplitude, and the input pulse frequency. But because two of these (duration and amplitude) are held constant, the DC output of the integrator is a function of only the frequency.

Half-Monostable Multivibrators

The half-monostable (*Figure 13-4a*) is based on either a single inverter or buffer and an R-C timing network (R1 and C1). In *Figure 13-4a*, the inverter version is shown.

At the initial point, capacitor C1 is uncharged, so the input of the inverter is at ground potential (see *Figure 13-4b*). This is interpreted as a LOW on the input; so the output is HIGH. The inverter sees a LOW whenever the voltage drops below some low value shown in *Figure 13-4b* as the "Low Input Voltage," and does not need to be exactly zero.

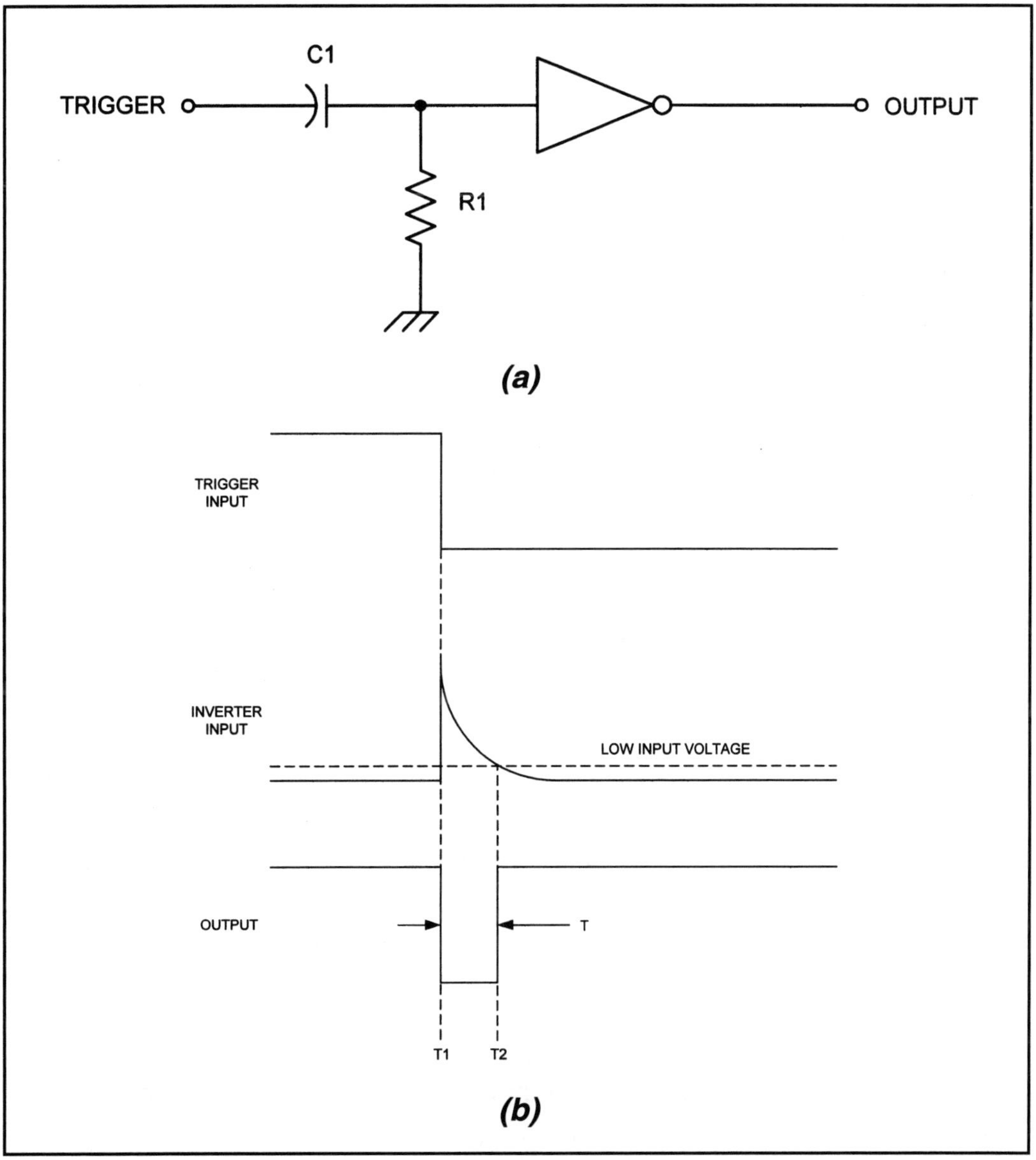

***Figure 13-4**. Half-monostable: a. circuit; b. timing diagram.*

When the circuit is triggered by a pulse that is longer than the output pulse duration, the capacitor C1 suddenly charges then discharges through R1 at a rate determined by the R1C1 time constant. The output duration is given approximately by:

$$T = 0.8\ R1\ C1$$

— where:

T is the output duration in second.
R1 is in ohms.
C1 is in farads.

The circuit shown in *Figure 13-4a* produces a negative-going output pulse for a positive-going input trigger pulse. The opposite situation (positive-going output for negative-going input) can be produced by lifting the ground end of resistor R1 and connecting it to the V+ DC power supply.

The same circuit can be implemented using buffer devices rather than inverters. If the inverter in *Figure 13-4a* is replaced with a buffer, then the positive-going input trigger pulse will produce a positive-going output pulse, exactly the opposite of the inverter. Similarly, if the ground end of R1 is connected to V+ in the buffer version, the output pulse is negative-going for a negative-going trigger pulse.

Operational Amplifier MMVs

Figure 13-5a shows the circuit of an MMV based on the operational amplifier device. Its timing waveform is shown in *Figure 13-5b*. For a moment, let's concentrate on the timing waveform, keeping in mind the definition of a one-shot. The MMV has only one stable output state. In this case, the stable state is at potential $+V_Z$. At time T1, the MMV is triggered, so the output snaps to the unstable state ($-V_Z$). Since the output has been HIGH, capacitor C2 will have to be charged to a positive voltage. This voltage would be $+V_Z$, except that diode D2 clamps the voltage to about 0.7 volts. When the output snaps LOW, however, the output voltage becomes $-V_Z$, so the capacitor will begin to charge toward that potential. But the voltage will not reach that value, for the noninverting input of the operational amplifier is biased to some fraction, ß, of the output voltage, as set by resistor network R7/R8. When the capacitor voltage exceeds $ßV_O$, the output of the operational amplifier will snap back to the HIGH stable state.

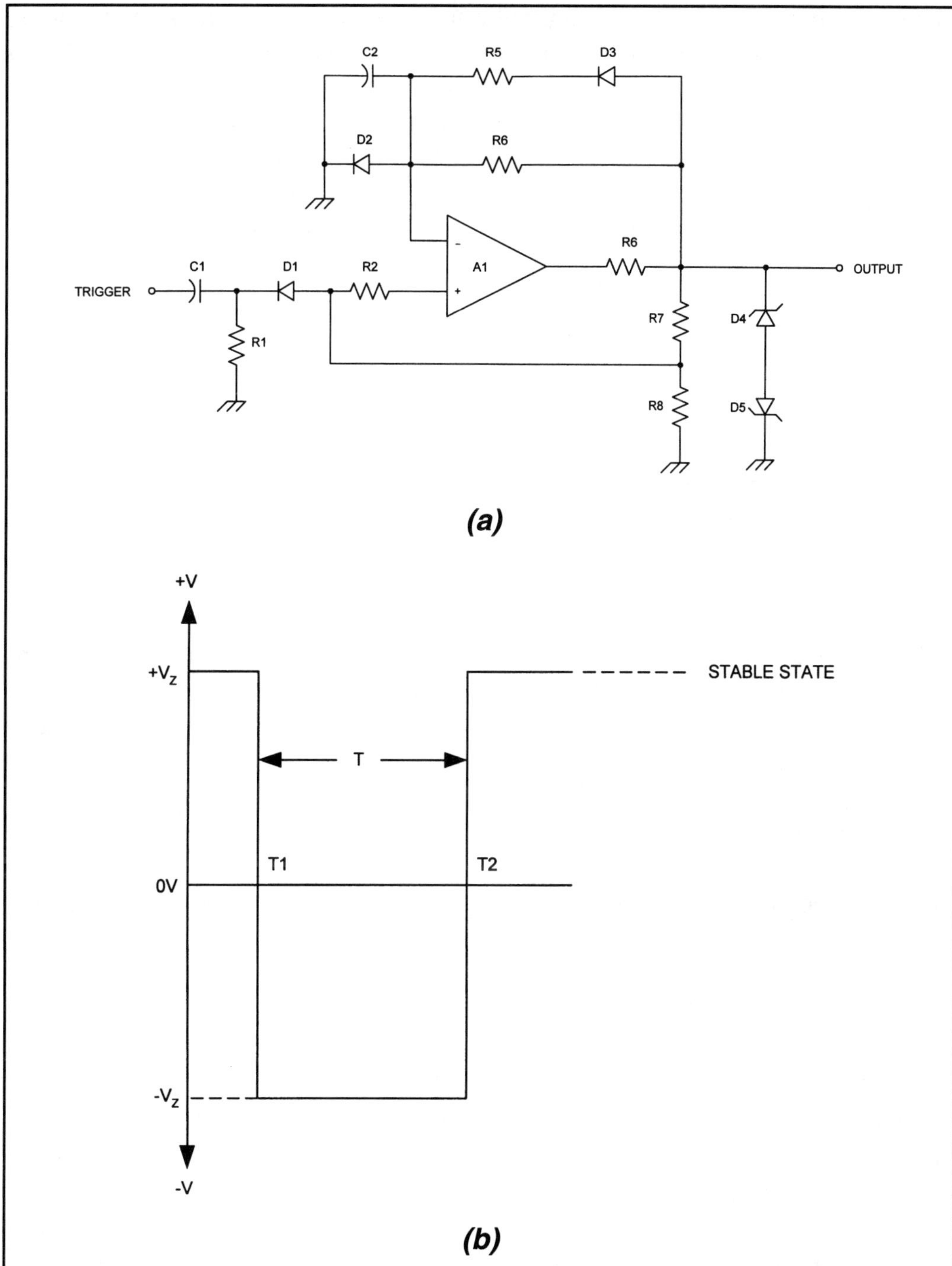

Figure 13-5. *a. Op-amp one-shot; b. timing diagram.*

The period immediately after the output returns to the stable state is called the *refractory period*, and the circuit will not respond to further trigger inputs during this period. Unless something is done, the capacitor will have to discharge under the influence of $-V_O$, R4 and C2, and that takes a long time. We can shorten the refractory period by placing diode D3 and resistor R5 in the circuit. If R5 >> R4, then the discharge cycle is a lot faster than the charge cycle. Diode D3 conducts only when the output voltage is positive. The duration of the output state is approximately:

$$T = R4\, C2\, Ln\left[\frac{1}{1-\beta}\right]$$

— where:

T is in seconds.
R4 is in ohms.
C2 is in farads.
Ln denotes the natural logarithms.
β is the feedback factor R8/(R7 + R8).

The equation is derived from the capacitor charging equation, which shows the time required to charge the capacitor (C2) from +0.7 volts to voltage βV_O. We can simplify the equation by assuming that R7 = R8 = R, then β = 0.5, so the equation becomes:

T = 0.69 R4 C2

This equation is often seen, and is the standard timing equation for op-amp based MMVs.

LM-555 MMVs

The LM-555 integrated circuit is a timing device that can be used as either an astable or a monostable multivibrator. *Figure 13-6a* shows the pinouts of the LM-555 device, while *Figure 13-6b* shows the circuit needed to make the LM-555 operate as an MMV. The duration of the output pulse is:

T = 1.1 R1 C1

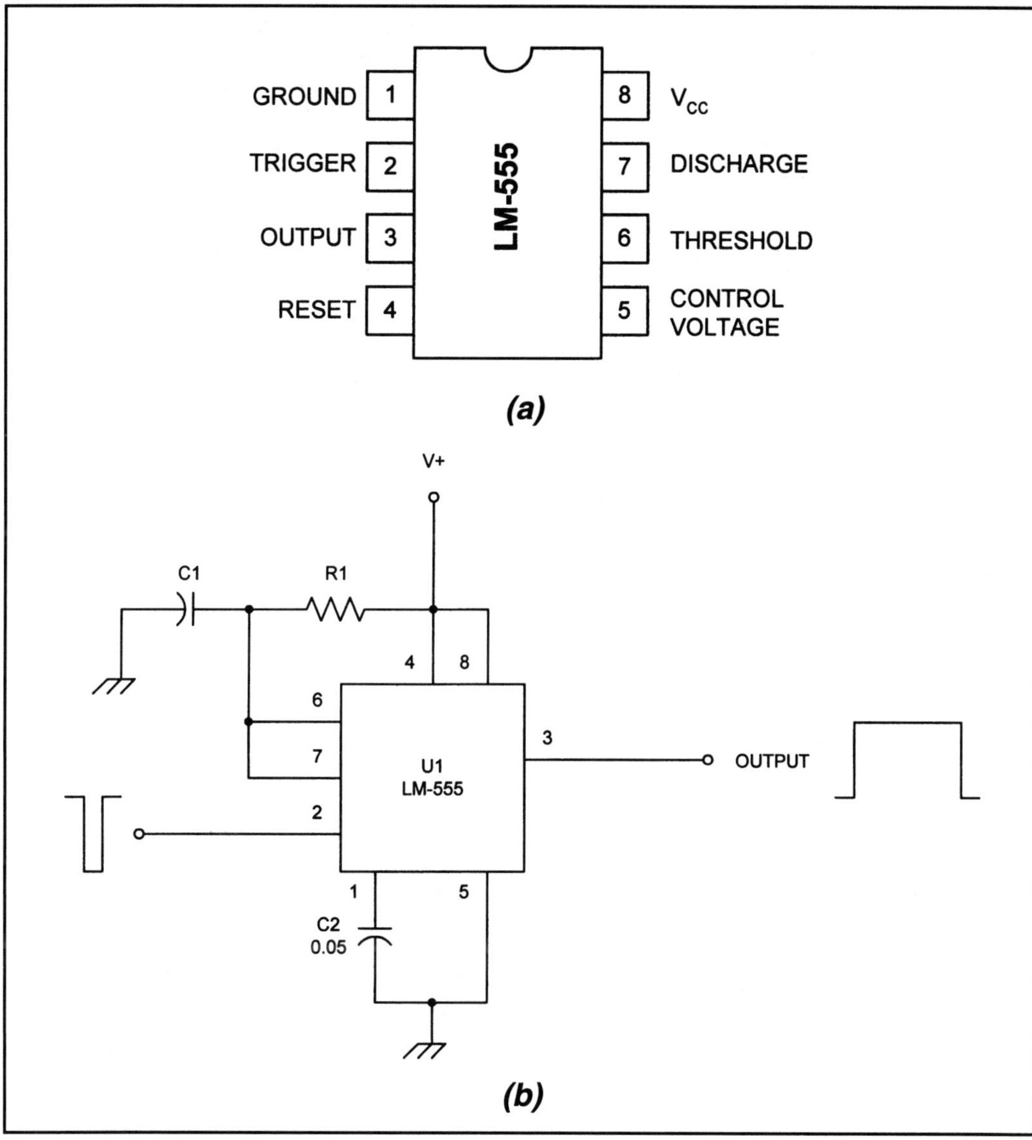

***Figure 13-6**. a. LM-555 pinouts; b. LM-555 one-shot circuit.*

The output will sink or source up to 200 mA of current, so it can be used for some relatively high current operations such as driving a relay or lamp. In this circuit, the stable state is LOW, or at ground potential. The unstable state, during which the pulse is active, is a positive voltage, the value of which depends on the value of DC power supply voltage (+5 to +18VDC).

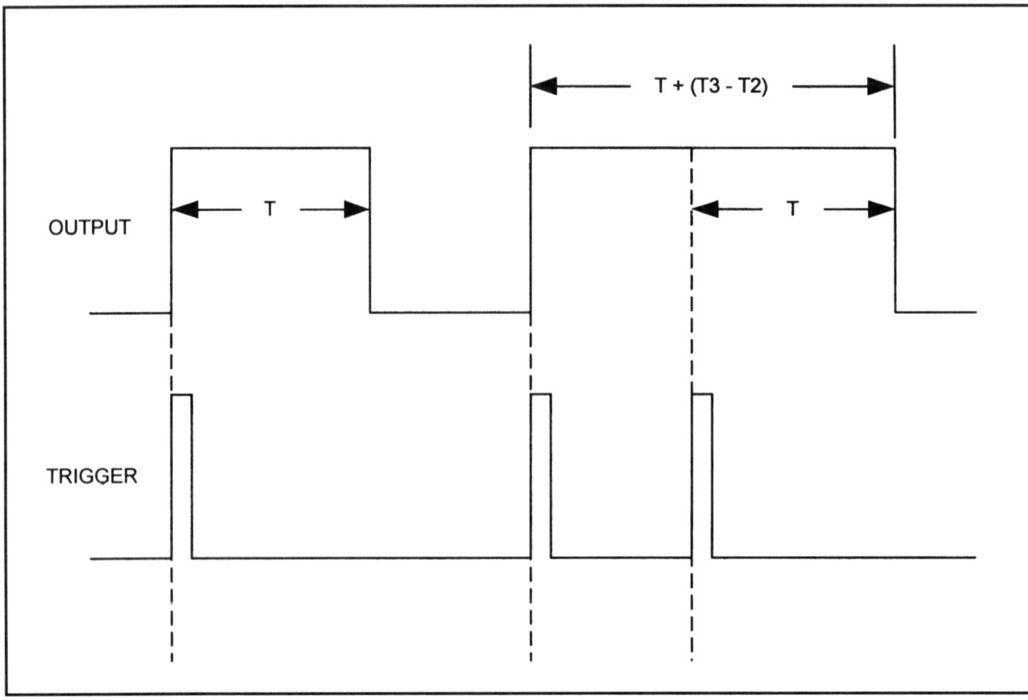

***Figure 13-7**. Timing diagram.*

Retriggerable MMVs

The retriggerable MMV differs from the standard MMV in that additional input trigger pulses extend the duration by another T. Consider *Figure 13-7*. At T1 the input is triggered, so a normal output pulse of duration T is observed. It times out, and at T2 a second pulse is received. This also triggers an output pulse of duration T, and it would run its course if no additional pulses are received. But at time T3, another trigger pulse is received before T times-out. As a result, the MMV is *retriggered*, producing another duration T output. The total duration is not 2T, but rather it is T plus the unexpired portion of the first pulse, or T + (T3-T2).

An application of the retriggerable MMV (or RMMV) is shown in *Figure 13-8*. This alarm circuit depends on an input signal that keeps the RMMV in the unstable state. If the RMMV ever times out, then the alarm circuit driven by its output is activated.

One place where this scheme is seen is in burglar and entry alarms that use a photoelectric sensor. It is used to foil people who would shine a flashlight into the sensor to fool it into believing that the source can still be seen (which would be interrupted given an intruder). The light source in this case is chopped, producing a series of light pulses. The sensor output, therefore, is a series of electrical pulses. As long as the pulses keep the RMMV triggered, then the alarm is not triggered. But if the pulses stop coming, then as soon as the RMMV times out, the alarm is sounded. I've also seen this circuit idea used in medical bedside monitor alarms. An electrocardiograph (ECG) signal or respiration signal is used to keep the RMMV triggered. But if the signal stops, indicating that the patient's heart has stopped beating or the patient has stopped breathing, then the RMMV times out and the alarm is sounded ... hopefully in time.

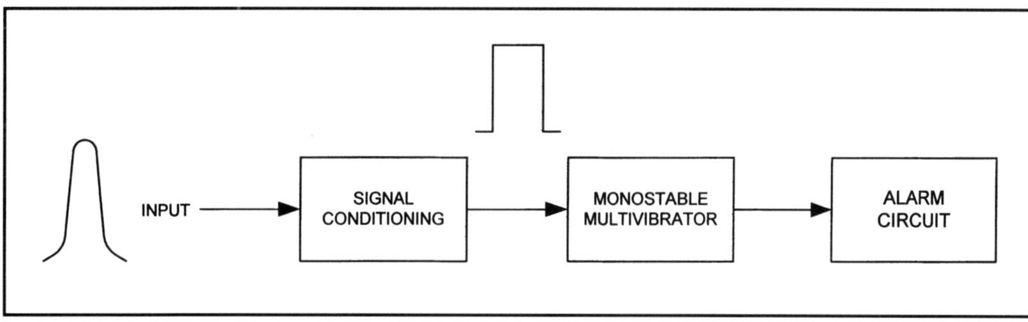

***Figure 13-8**. Alarm circuit.*

Chapter 14

Digital Clocks and Astable Multivibrators

An *astable multivibrator* is a digital circuit that has no stable state. It will alternate back and forth between HIGH and LOW output states as long as the circuit is turned on. This is in contrast to the monostable multivibrator (one-shot), which has one stable state and one unstable state. The one-shot remains in the stable state until triggered, then it snaps to the unstable state for a specified period of time. The astable multivibrator oscillates back and forth between the two unstable states, producing a square wave output (*Figure 14-1*).

Generating Square Waves from Sine Waves

Schmidt Trigger Approach. A chip called the *Schmidt trigger* can be used to generate square waves from a sine wave input. It is sometimes preferable to use an analog oscillator to generate the signal because these circuits are often easier to stabilize frequency wise. But such a circuit will produce a sine wave output that must be converted. The circuit for the converter is shown in *Figure 14-2a*, while the operation of the Schmidt trigger is shown in *Figure 14-2b*. The operation of the Schmidt trigger follows this simple rule: the output will snap HIGH when a positive-going input signal crosses a certain threshold. It will snap LOW again when the input signal subsequently crosses a lower threshold in a negative-going direction. Thus,

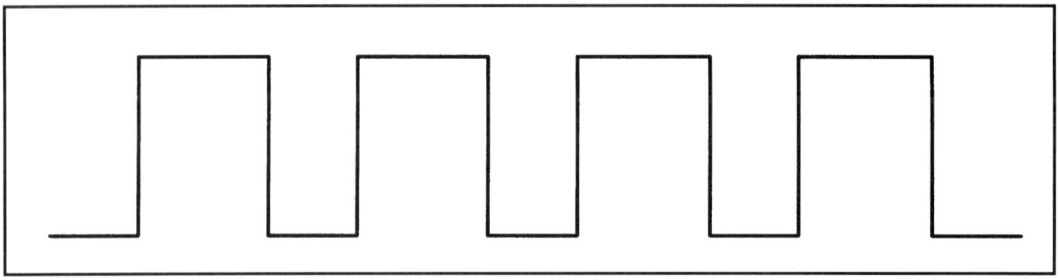

***Figure 14-1**. Square waveform.*

when a signal from an analog oscillator crosses +1.7 volts (*Figure 14-2*) in the positive-going direction, the Schmidt trigger output snaps HIGH. It will drop LOW again when the input signals drops to 0.9 volts in a negative-going direction. The output of the Schmidt trigger is a square wave with the frequency of the sine wave input signal.

Comparator Approach. A voltage comparator is an extremely high-gain differential amplifier. An example is an operational amplifier without feedback. This type of amplifier will saturate with only a very tiny voltage differential between the two inputs. The output is zero only when the input voltages are equal. If the inverting input is slightly more positive than the noninverting input, then the output saturates to the V- power supply rail. Alternatively, when the inverting input is less positive than the noninverting input, the output snaps to the V+ power supply rail. *Figure 14-3* shows this effect. Some IC comparators are designed to operate from a single-polarity DC power supply (such as LM-311). If the output of a sine wave oscillator is applied to one input of a comparator, and the other input is grounded, then the output will snap HIGH and LOW at the zero crossings.

TTL and CMOS Clock and Oscillator Circuits

A TTL or CMOS oscillator can be built using gates or inverters, with either piezoelectric crystals or resistor-capacitor (R-C) networks to control the operating frequency. A small collection of TTL oscillators is shown in *Figure 14-4*. These devices are built using either TTL inverters or TTL NAND or NOR gates connected as an inverter (inverter connection occurs on both NAND and NOR if both inputs are tied together). The first circuit, in *Figure 14-4a*, is R-C timed; that is, the oscillating frequency is determined by capacitor C1 and resistors R1 and R2. In this particular configuration, the resistance is adjustable in order to continuously vary the operating

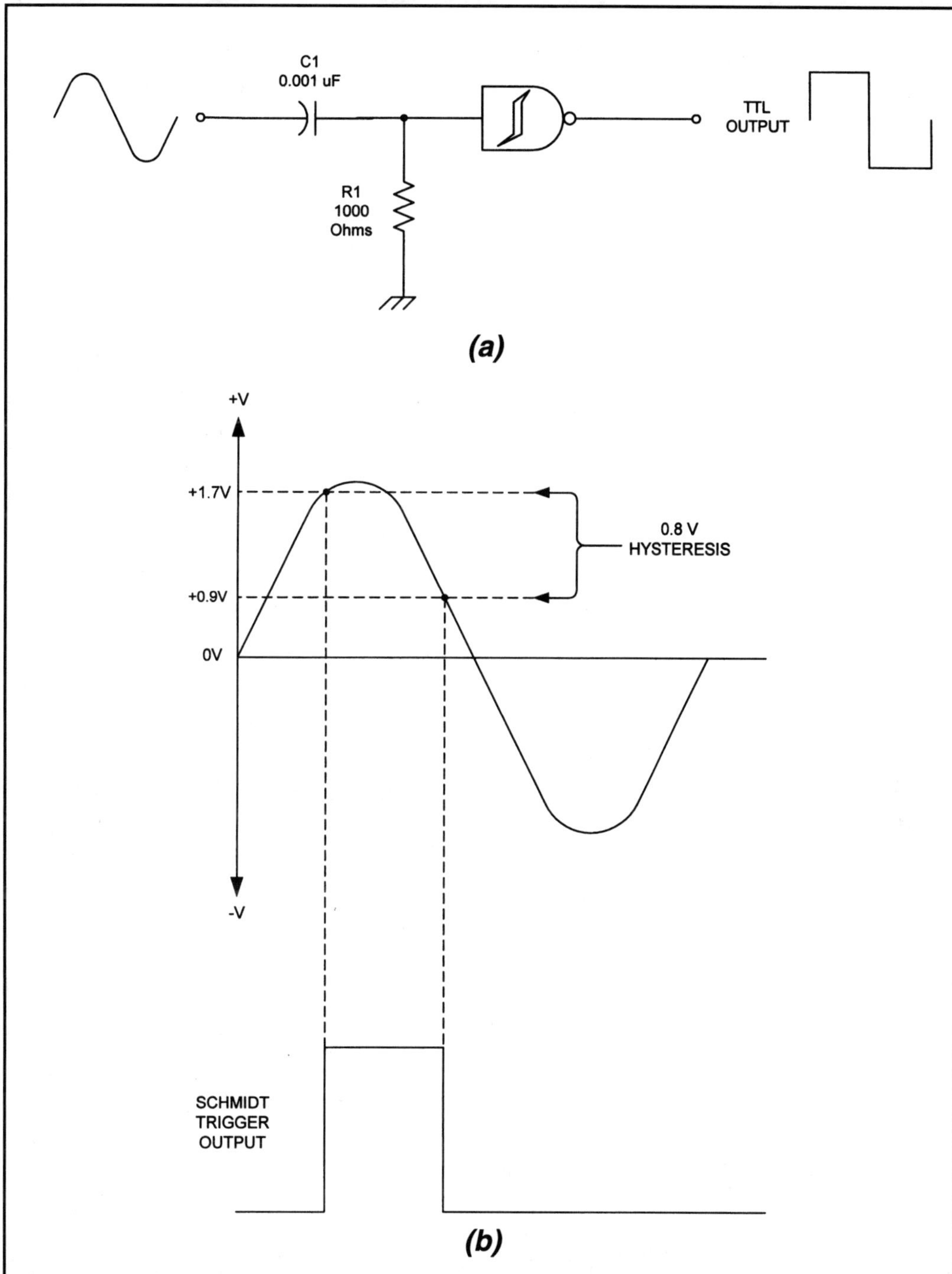

Figure 14-2. *a. Schmidt trigger squarer; b. timing diagram.*

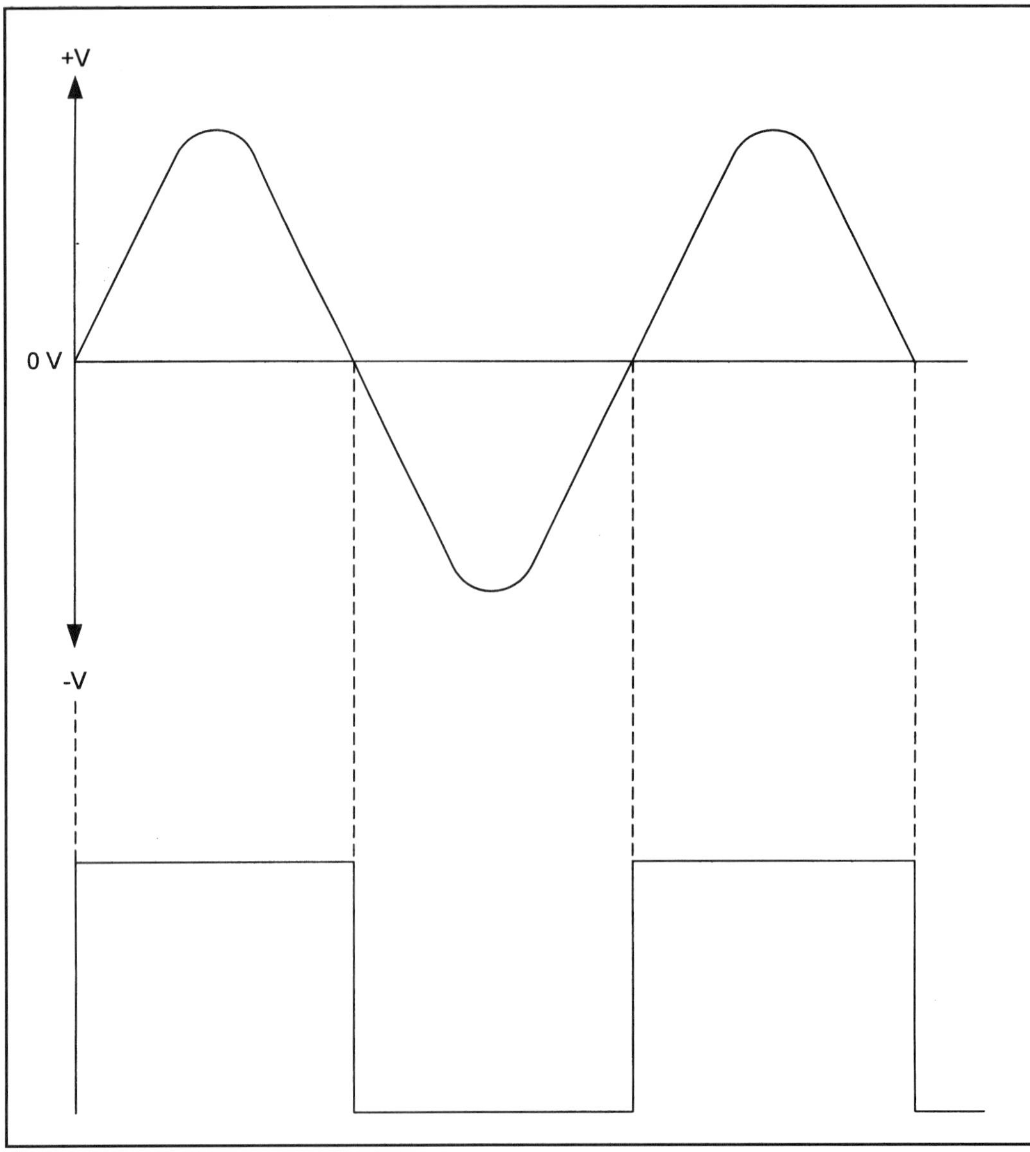

***Figure 14-3**. Using a zero-crossing detector to make square waves from sine waves.*

frequency over a small range. If only a single fixed frequency is needed, then replace potentiometer R1 and resistor R2 with a single fixed resistor.

A disadvantage of any RC-timed oscillator is that the operating frequency is neither stable nor accurate. These problems are partially overcome using piezoelectric crys-

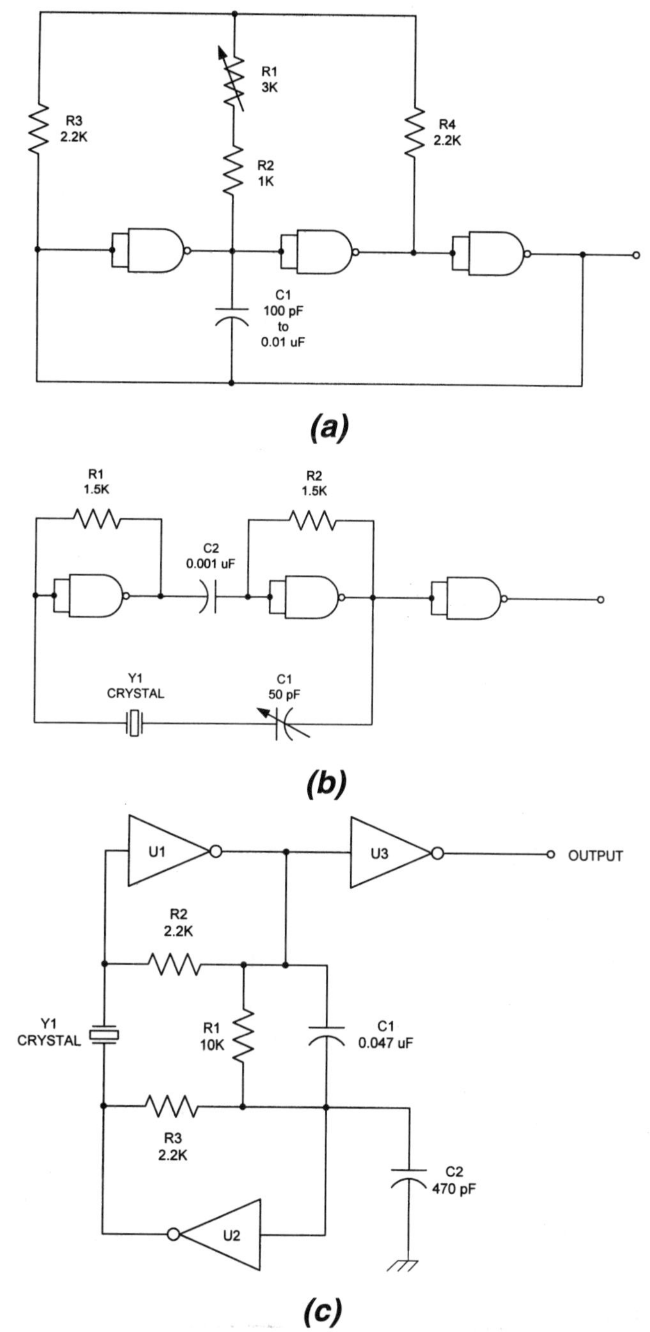

***Figure 14-4**. a. R-C timed TTL astable; b. crystal timed version; c. CMOS crystal oscillator.*

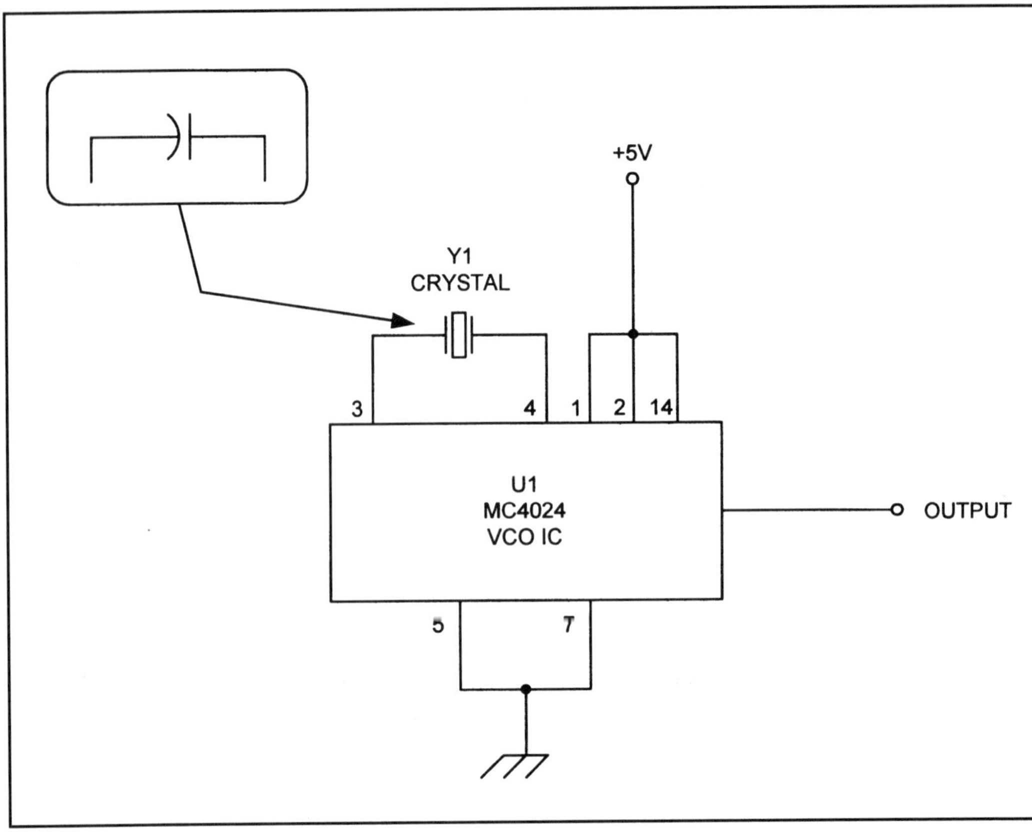

Figure 14-5. *MC-4024 crystal voltage controlled oscillator.*

tals, as in *Figure 14-4b* and *Figure 14-4c*. The circuit in *Figure 14-4b* is based on a standard TTL NAND gate. For example, the 7400 device contains four independent NAND gates, of which two are used for the oscillator and a third acts as a buffer stage. All three NAND gates are connected as inverters. The two gates (A and B) used in the oscillator portion of the circuit are each biased with a 1.5 kohm resistor, and are isolated from each other by coupling capacitor C2. The operating frequency is set by crystal Y1, and may be varied slightly with capacitor C1.

The version shown in *Figure 14-4c* is similar, and is based on TTL inverters. These stages can be made using inverter-connected NAND or NOR gates, or selected as part of a hex inverter. Again, one stage is used as an output buffer, while the oscillating stages are self-biased by a feedback resistor.

Special IC Oscillators

There are several TTL oscillators available on the market. In *Figure 14-5*, we see the MC-4024P device. It is a *voltage controlled oscillator* (VCO), meaning that the output frequency is at least partially determined by the value of a DC control voltage applied to a special pin. Note that this device should not be confused with the CMOS 4024 device, which is a binary counter. Rather, the MC-4024P device contains two linear VCOs with a TTL compatible output signal. In this circuit, only one side of the device is used. A second, completely independent oscillator could be built with the other half.

There are two approaches to setting the overall operating frequency (other than controlling the input voltage): *capacitor* or *crystal*. In most cases, the capacitor is used and will have a value of approximately 300/F picofarads, where F is the frequency in megahertz. For frequencies above 2,000 kHz, however, you may also elect to use a crystal to set the operating frequency (both methods are shown in *Figure 14-5*).

The circuit in *Figure 14-5* has the control voltage input pin (number 2) connected to +5VDC for fixed frequency operation. If we want to make the frequency variable, then we would use a circuit such as *Figure 14-6*. In this case, a potentiometer and

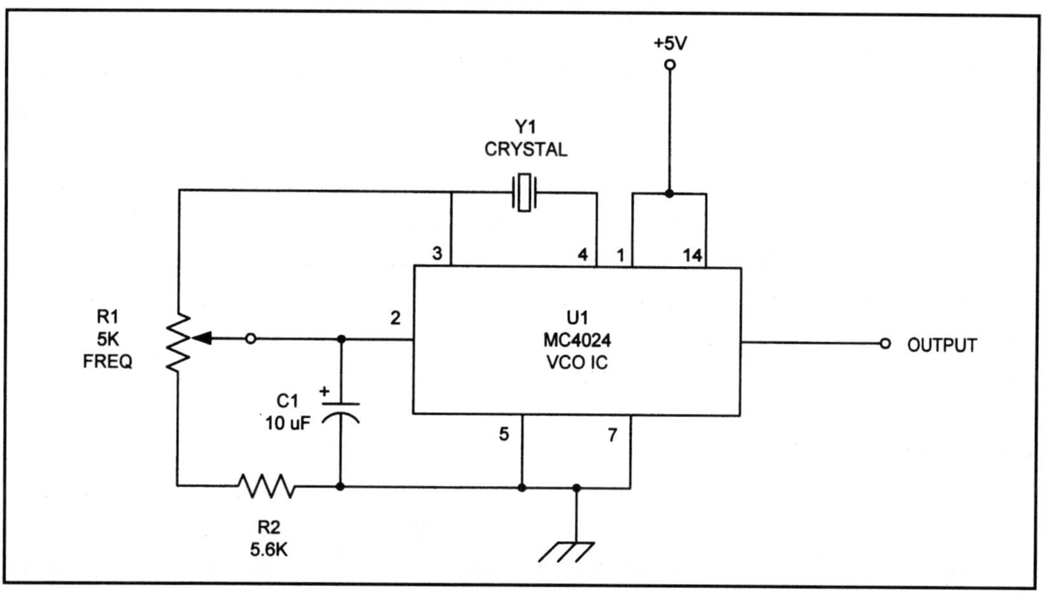

Figure 14-6. *Crystal MC-4024 oscillator.*

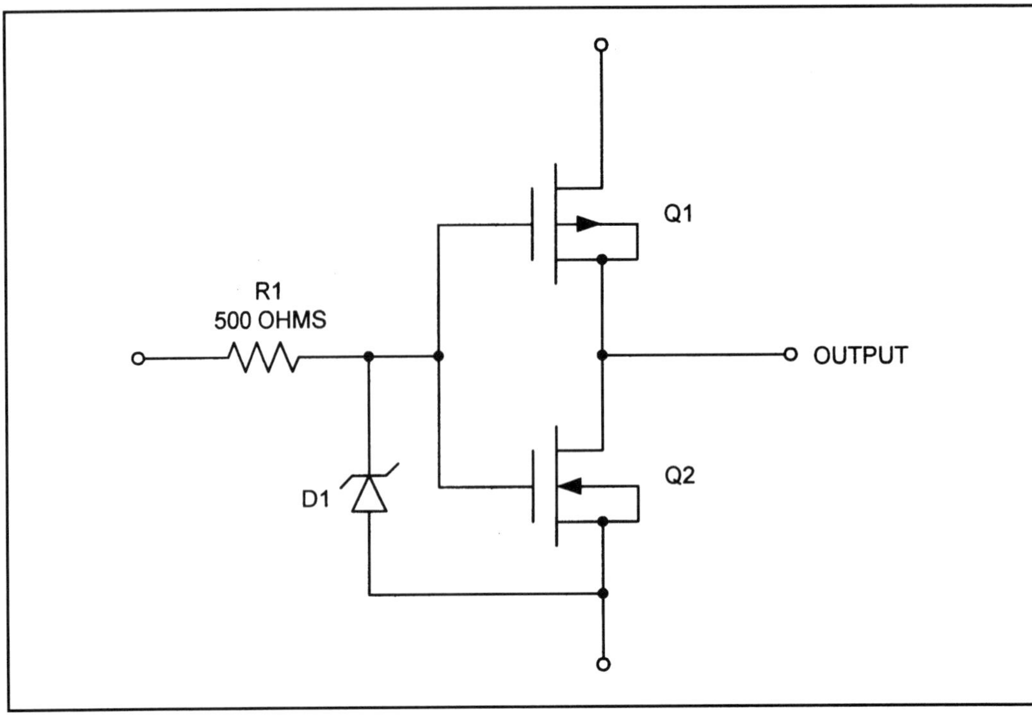

***Figure 14-7**. CMOS inverter.*

fixed resistor connected in a voltage divider circuit sets the control input voltage at some potential less than +5 volts — with subsequent pulling of the operating frequency. When a capacitor is used in place of crystal Y1, the frequency can be pulled over a 3:1 range. The control input is grounded for AC circuits by capacitor C1, which also serves to stabilize the voltage (also the frequency).

CMOS Clock Circuits. The next type of oscillator/clock circuit that we will consider is the CMOS digital logic oscillator. Like their TTL counterparts, the CMOS oscillator is often used as the block in digital electronic circuits and computers for timing and synchronization. There are some significant differences between TTL and CMOS, however.

Perhaps the biggest difference, and one that drives the other attributes, is that CMOS devices are made of metal oxide field effect transistors (MOSFETs) instead of the NPN and PNP bipolar transistors that are used in TTL devices. An implication of this fact is that CMOS devices draw considerably less current than TTL devices (microamperes instead of milliamperes). However, the lower current is often at the expense of operating speed. Both fundamental oscillation frequency and pulse rise times suffer in some CMOS circuits compared to TTL equivalents.

Figure 14-7 shows the basic CMOS inverter circuit used to make our oscillator circuits. The basic inverter is representative of the now-obsolete "A series" CMOS devices. It consists of a pair of MOSFET transistors with their drain/source paths in series and the gates in parallel. Device Q1 is an N-channel MOSFET, while Q2 is a P-channel MOSFET. These devices operate in opposite polarities from one another. A HIGH applied to the common parallel-connected fate will cause one of these transistors to saturate on and the other to pinch off.

The newer B series CMOS device is the type of device currently on the market; although if one looks, one will find some A series available. There is a resistor in series with a gate, and at least one diode shunting the gate. The reason for these components is electrostatic discharge (ESD) protection. Normal A series devices have a gate breakdown voltage of around 80 volts. Since static buildup on tools and clothing can easily reach several thousand volts, you can easily damage an A series device. Even though the B series devices are ESD protected by the resistor and zener diode, one must still exercise caution and follow ESD precautions when handling the devices.

Most CMOS devices have part numbers in the 4000 and 4500 series. Some are listed bare (such as 4024B and 4528), while others have a prefix such as CD-4xxx or MC-144xxx. The "A" or "B" suffix on the part number indicates the series. In general, if the device is new production (and not fifteen years old), then lack of a letter suffix still denotes a B series or later device.

There are other differences between the A series and B series devices. The B series offer faster rise and fall times for square wave pulses, and will drive larger loads than their A series cousins. Although there may be some applications where the characteristics of the A series provide some design advantage, it is generally recommended that B series devices be used.

The transistors in the CMOS device can be modeled as a pair of gate-controlled resistors in series. In *Figure 14-7*, the channel resistance of Q1 is represented by R1.

CMOS devices operate from two DC power supplies, V+ and V-, both of which can be roughly 4 to 18 volts, although sometimes V- is 0V. It is common to find V+ set to +5VDC to make the device compatible with TTL devices (which require +5VDC power supply). In cases where TTL compatibility is an issue, the V supply is set to zero volts. It is worth noting that the published operating specifications and parameters of the CMOS devices assume a ±10VDC power supply. Operation at other potentials may differ somewhat from the published version.

Let's examine what happens when a signal is applied to an input gate. Assume that V+ = 5V (which also represents logical HIGH), and V- is grounded, or 0V (which represents logical LOW). When the input is LOW, then the output is HIGH. This means that resistor R1 has a low value (usually less than 2,000 ohms) while R2 is very high (>1 megohm). In this case, the output terminal is effectively connected to V+, so the output is HIGH. Alternatively, when the input is HIGH, the resistances reverse: R1 becomes very high and R2 becomes low. In this case, the output terminal is effectively connected to ground.

The reason for the very low current drain of the CMOS device is that the V- to V+ DC path is always a very high resistance in series with a low resistance, except for the instant of transition between HIGH and LOW, or vice versa. There will be a brief but usually insignificant spike in current demand at the instant of transition.

Our first CMOS clock/oscillator circuit is shown in *Figure 14-8*. The circuit is in *Figure 14-8a*, while the timing diagram is in *Figure 14-8b*. The active device is a CMOS Schmidt trigger (four such triggers come in the 4093B device). A Schmidt trigger, you will recall, is a special circuit that operates or toggles the output only under specific input voltage conditions. Assuming that the power supply is +5 volts and grounded, the inputs operate on positive-going signals that pass through +2.9 volts and negative-going signals that pass through +2.3 volts. The device must see a positive-going input prior to the negative-going threshold crossing, or it will not operate.

The device shown in *Figure 14-8a* is a two-input NAND gate, so we can either tie the two inputs together or use one of the inputs as a control element. When the switch is open, the input is at V+, so the device will operate normally. But when the switch is closed, the input is grounded so the device will sit there dormant with the

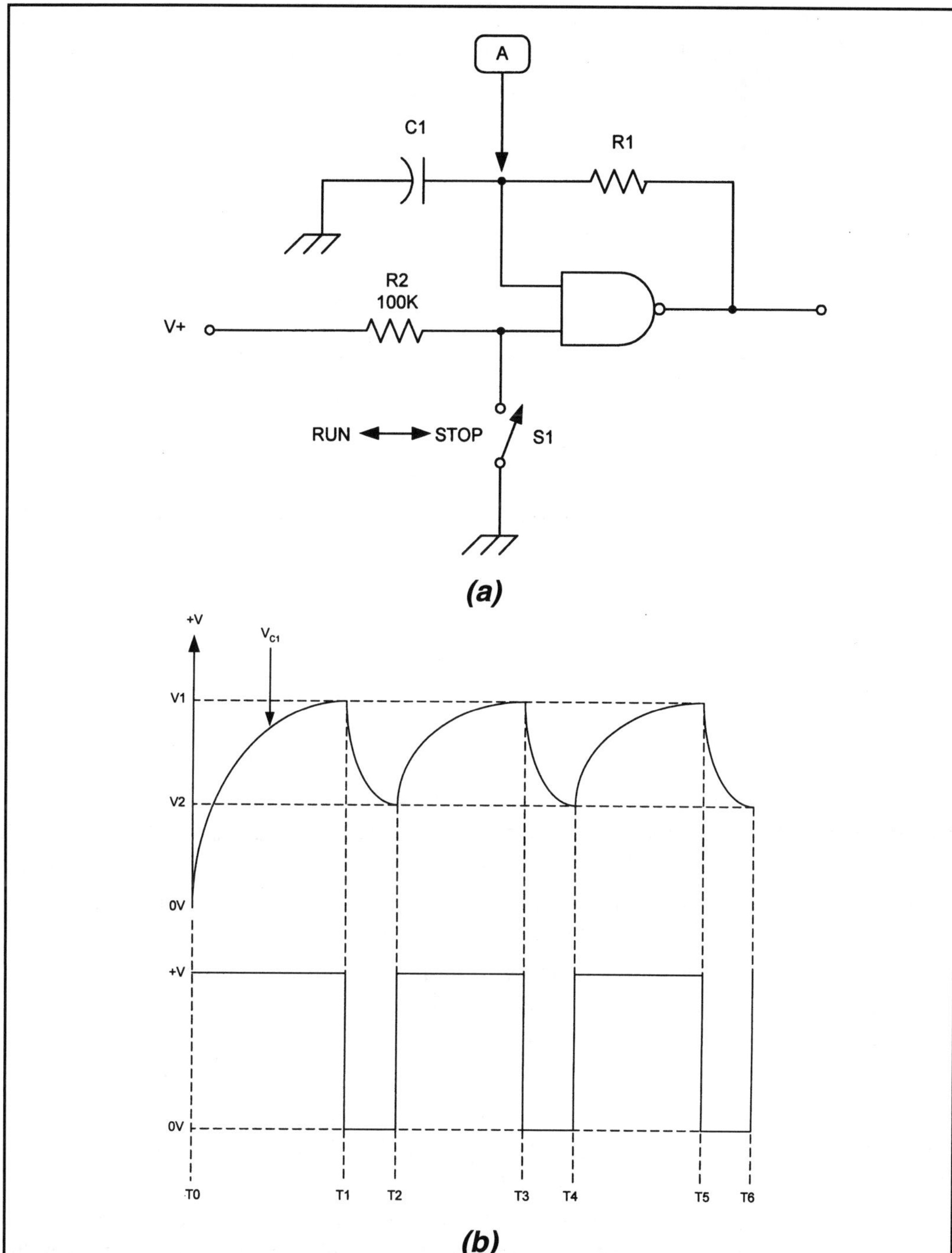

Figure 14-8. *a. Astable MMV with run/stop switch; b. timing diagram.*

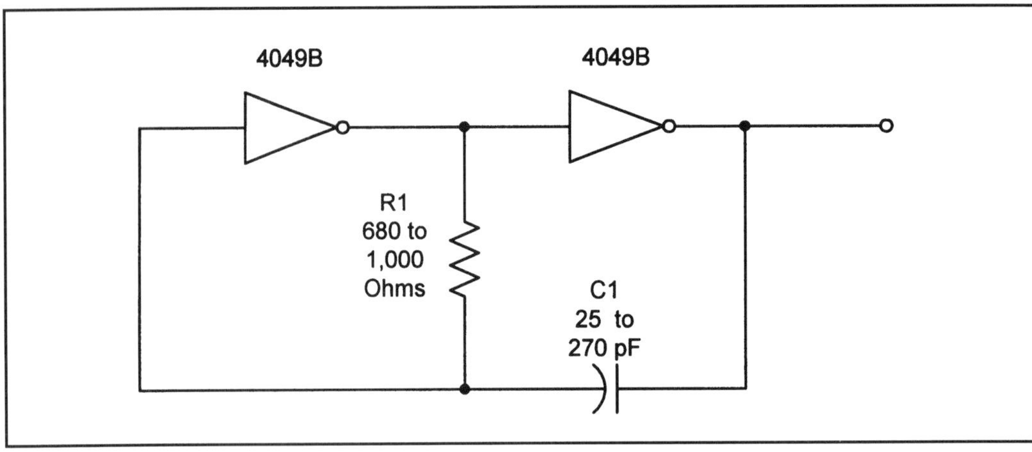

***Figure 14-9**. CMOS astable oscillator.*

output in the HIGH state. For our description of operation, assume that the switch is open.

At turn-on, capacitor C1 is discharged, so it has a potential of 0 volts at point "A," which is ground potential. This makes the input LOW. The output is therefore HIGH at this instant (see time T0 in *Figure 14-8b*). The voltage across the capacitor begins rising as C1 charges from the output potential through R1. The voltage, VC1 in *Figure 14-8b*, will eventually reach the positive-going threshold, V1. At this time, the input sees a HIGH, so it will cause the output to drop LOW. This change of condition removes the charge source from R1C1, so C1 will begin to discharge through R1 and the output terminal of the gate. When the voltage decreases to V2 (the lower or negative-going threshold), the output snaps HIGH and the cycle repeats itself.

A second CMOS oscillator (this one a little more common) is shown in *Figure 14-9*. This circuit can be made using a CMOS inverter, or either NAND or NOR gates connected as inverters (or with both inputs tied together). The operating frequency of the circuit is set by the resistor-capacitor combination according to the approximation $F_{HZ} = 10^6/(2.2\ R1C1)$, where F is in hertz, R1 is in ohms and C1 is in microfarads (µF). The resistor (R1) can have any value from about 3,300 ohms to several megohms, and the capacitor any value from the values shown up to about 10 µF.

The output waveform of this device is a little unsymmetrical, especially when B series CMOS devices are used. The problem with the B series devices is that the

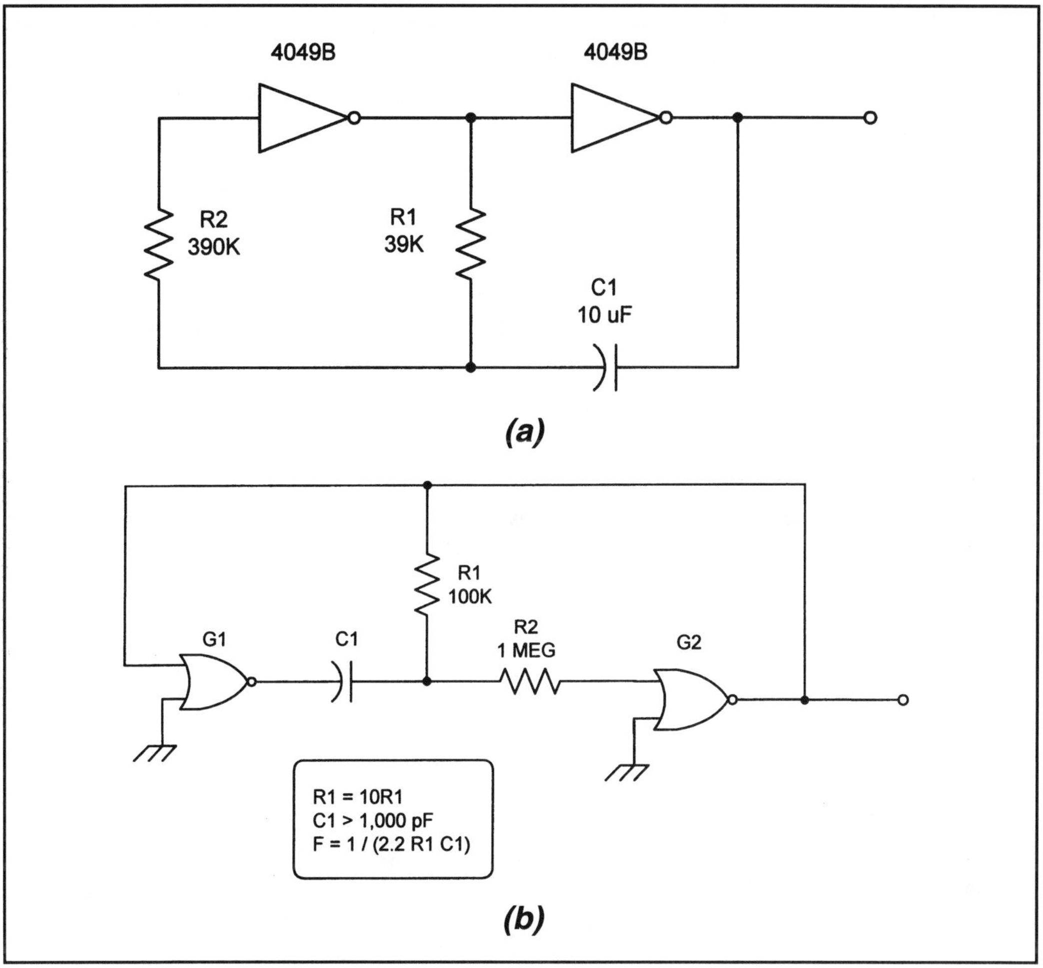

***Figure 14-10**. a. CMOS astable oscillator; b. astable oscillator.*

internal diode protection disturbs the situation, so some means must be provided to isolate the R1C1 combination from the protected input. *Figure 14-10a* and *Figure 14-10b* show how this is done. Resistor R2 in both cases isolates the RC network from the diode-protected input, and cures the asymmetry problem in the waveform. These circuits are generally preferred over those of *Figure 14-9.*

A pair of voltage controlled oscillators (VCO) are shown in *Figure 14-11a* and *Figure 14-11b.* These circuits produce an output frequency that is partially dependent upon the value of the control voltage applied to the input. In *Figure 14-11a*, this voltage is applied in a manner to bias the input of the inverter (and the junction of the

RC network) to some static value. The charge/discharge cycle that sets the operating frequency takes longer or shorter durations depending upon the applied voltage. The $V_{IN} = 0$ set frequency is found according to the formula given previously.

The circuit in *Figure 14-11b* works a little differently. In this case, a MOSFET transistor shunts the RC junction to ground, thereby affecting the charge/discharge timing. The control voltage is applied to the gate input of the MOSFET transistor.

The 4046 device shown in *Figure 14-12* is an integrated circuit CMOS voltage-controlled oscillator. The chip is listed in the data sheet as a phase-locked loop, but it can be operated as a VCO. The control voltage input to pin number 9 is provided by a potentiometer or other source. The operating frequency is determined by C1 and R1, and to some extent by the optional resistor R3. The value of R1 can be anything from 3,000 to 1,000,000 ohms, while the capacitor may be anything from 100 pF to 10 µF. It is possible to make an oscillator that operates over a range of 1000:1 with the 4046 device.

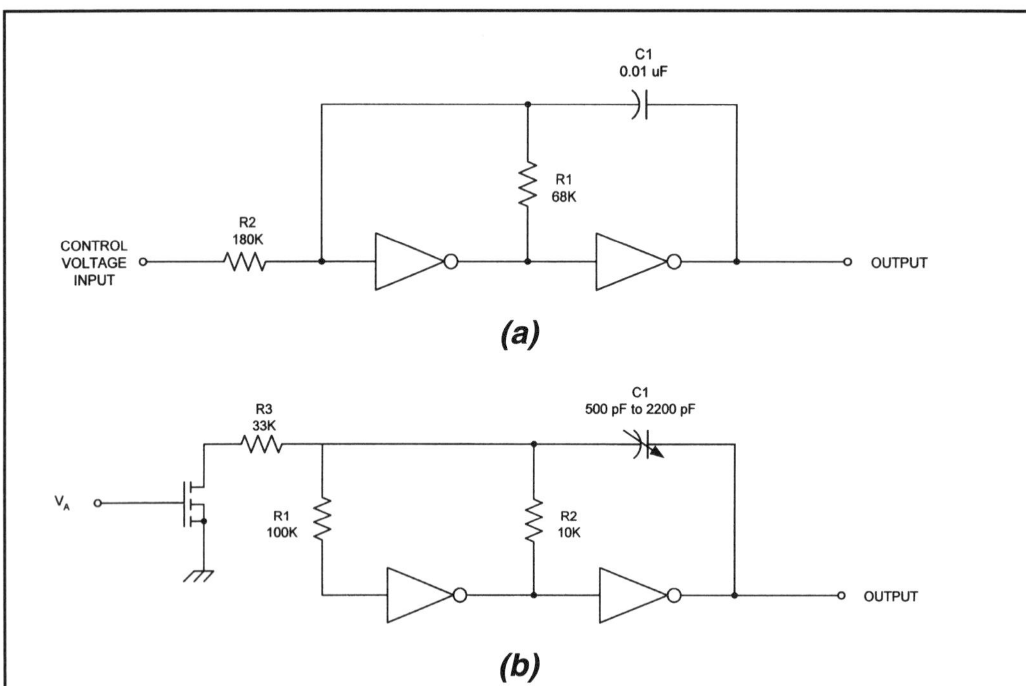

***Figure 14-11**. VCO circuits.*

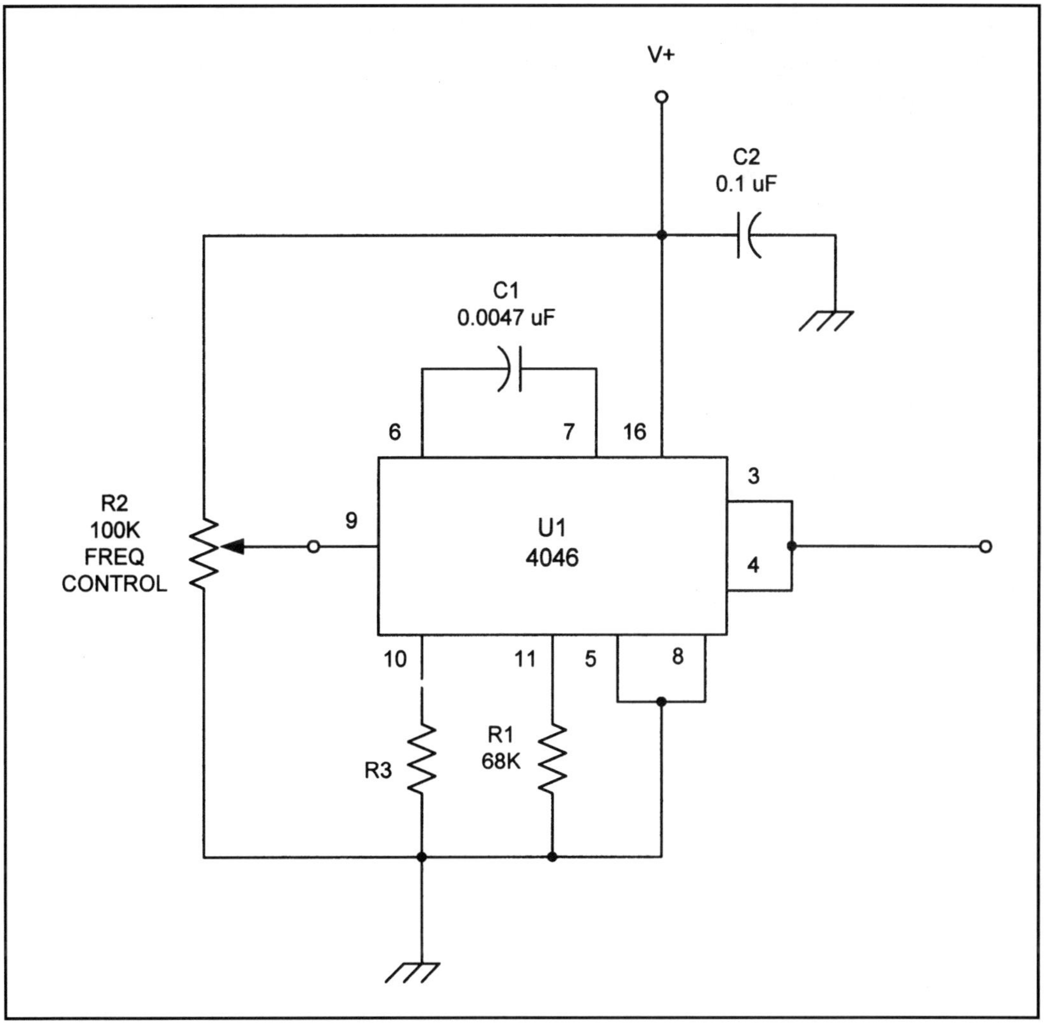

***Figure 14-12**. CMOS oscillator.*

The use of R3 varies the minimum and maximum operating frequencies. Ordinarily, with R3 open circuited, the operating frequency will vary from very near DC to the maximum set by the R1C1 network, or about 1 MHz or so. If R3 is used, however, the operating frequency range changes. There will be a minimum and a maximum that differ from the above rule.

The final example is the crystal oscillator shown in *Figure 14-13*. This circuit uses two sections of a CMOS NOR gate, both of which are inverter connected. One section operates at the oscillator, while the other is a buffer stage that serves to isolate the oscillator output from the load. The crystal oscillator is an exception to the digital types of circuit shown elsewhere. The reason is that the NOR gate is biased into a linear region of operation by R1, so it acts as an amplifier. Gains up to 10,000 are possible with buffered B series devices. The operating frequency is set by the piezoelectric crystal resonator (Y1), operated in its parallel resonant mode. The RC components are selected to provide a 180 degree phase shift at the crystal's parallel resonant frequency. This circuit will provide good stability and accuracy of output frequency.

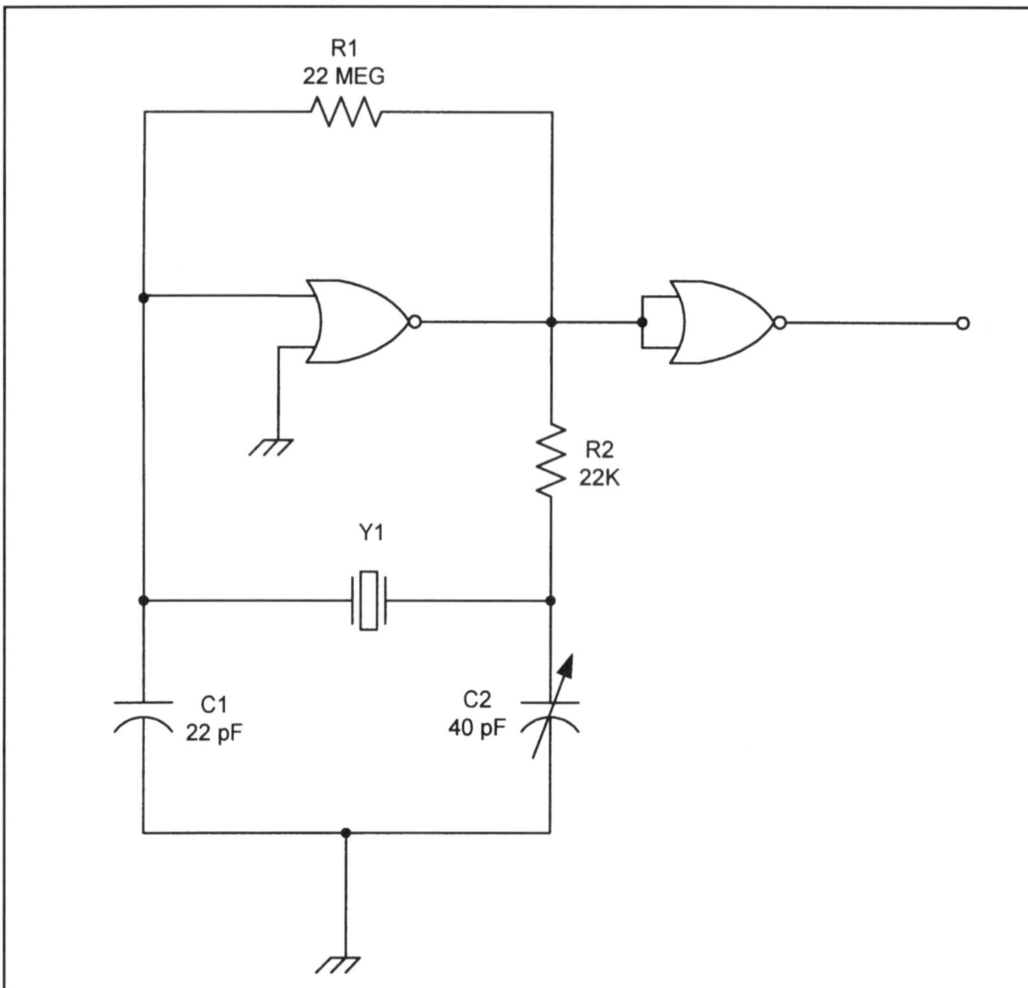

Figure 14-13. *CMOS crystal oscillator.*

Electrostatic Discharge (ESD) Damage

There is one major problem with CMOS and related logic devices: they are easily damaged by static electricity discharges. Almost everyone is familiar with static shocks. If you walk across a rug, or if you wear certain kinds of cloth (such as wool and many synthetics), then a static electric charge builds up on your body. Touch something that is grounded, and there will be a small but startling spark. The spark represents very little electrical energy and current, but has a potential up to several thousand volts. These potentials can blow a CMOS device. Be sure to follow normal CMOS protection methods when using these devices.

Chapter 15

Some Common Problems

There are two principal areas where digital equipment can have recurrent problems: the DC power supply and the temperature. These two considerations cause many different types of problems, including recurrent problems, "tough dog" intermittent problems, or both.

Any piece of digital equipment that contains substantial amounts of TTL devices will generate lots of heat. The high current requirements of any fast (or high frequency) logic family will mean a large amount of heat energy being dissipated. This is one argument for replacing lots of logic with software, wherever possible. Unfortunately, many devices, especially when used in combination with (or synchronized to) other circuits, become flaky when the operating temperature goes above a certain point. As a consequence, one must pay close attention to the overheating problems. One stand-alone minicomputer, for example, uses forced air blowers to keep the internal temperature down. The manufacturer has placed a hole in the cabinet at a point where the technician can insert an ordinary medical clinical thermometer. Although temperatures at other points are higher, this indicator point is useful because it allows easily-available and low-cost test devices (oral thermometers) to be used. The specification for the air temperature at that point in the cabinet is 102 degrees F, or 38.9 degrees C.

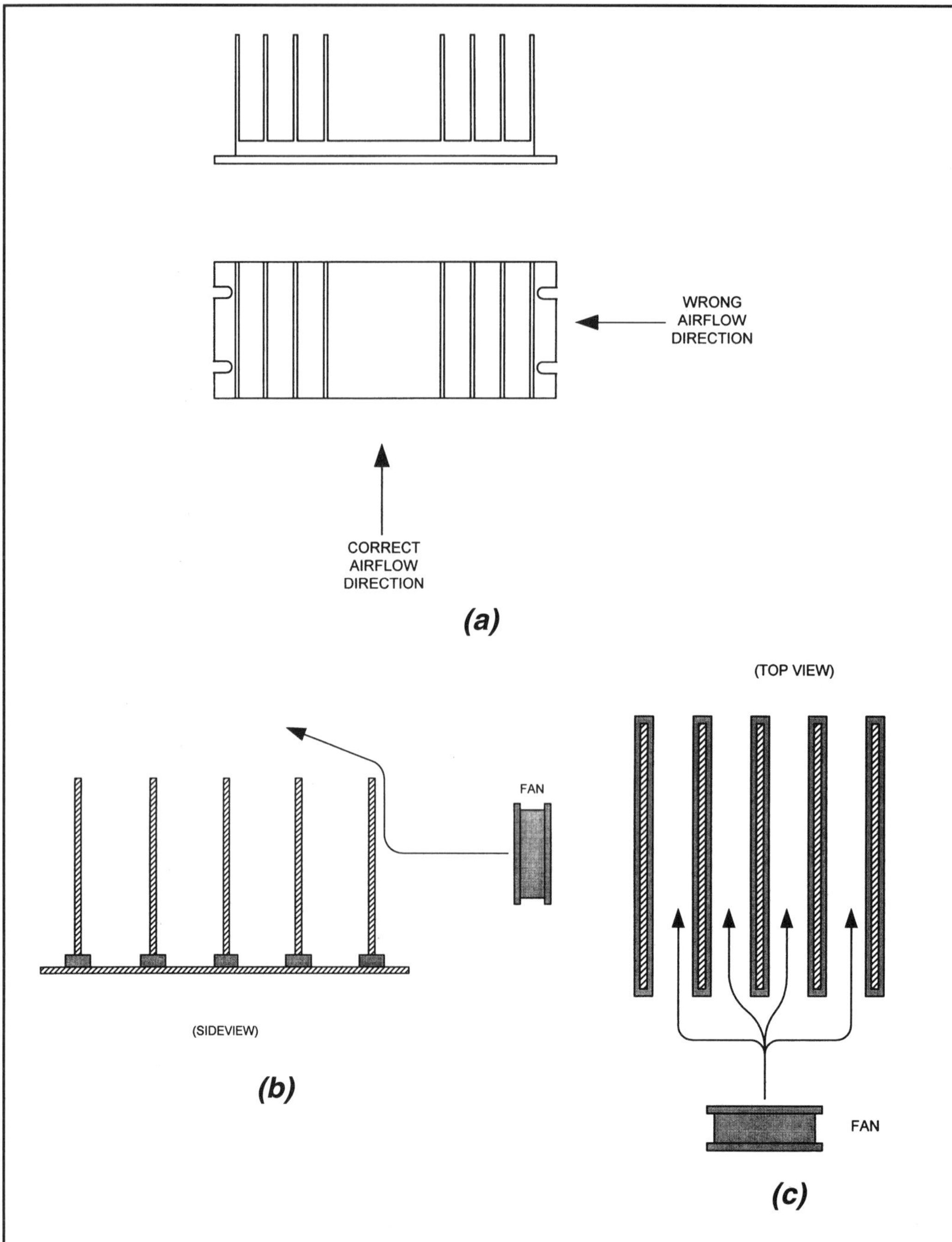

***Figure 15-1**. a. Correct air flow for heat sink; b. incorrect flow for plug-in boards; c. correct flow for plug-in boards.*

DC power supplies frequently encountered in larger digital systems are usually equipped with a thermal shutdown circuit. These circuits will turn off the power supply when the over-temperature condition persists for longer than a specified length of time. Also, series-pass and switching transistors tend to short-out the collector-to-emitter at temperatures over some point above 125 degrees C. They are even derated at temperatures above some that are lower than the upper limit (such as at 110 degrees C).

Temperature, therefore, can easily be a critical parameter in digital equipment. One might suppose that manufacturers would see to these matters in the original design of the beast. Usually, they do. But sometimes Murphy's law takes over and a device is designed with a deficiency. We also find instances where a user will operate the equipment at temperatures higher than those specified by the manufacturer.

The solution to heat problems is to:

1. Reduce the ambient temperature.
2. Increase the flow of air or coolant over the circuit to carry away heat.
3. Or do both.

In most cases, electronic equipment is cooled with forced air. Ventilation holes must be placed so that air blown across the affected circuitry can escape and carry the heat with it.

There are right and wrong ways to eliminate heat from equipment. *Figure 15-1* shows several possible situations. In *Figure 15-1a* is a finned heat sink. The heat sink should be physically placed so that air from the forced air fan or blower travels *between the fins*. If it travels perpendicular to the fins, then it will not see maximum exposed surface area. Blowing air across the maximum surface area tends to maximize the heat transfer.

A similar situation exists with respect to printed circuit boards inside of a cabinet. The fan or blower should be positioned to blow across the component side of the printed circuit boards. *Figure 15-1b* shows the incorrect way to position the fan, while *Figure 15-1c* shows the correct placement. Note that in *Figure 15-1c*, the fan should usually be placed to blow evenly across all of the boards. If one of the boards is known to run hotter than the others, however, the position of the fan can be altered to blow predominantly over that board. You may also need to add an additional fan or blower to cover that particular board. There are also times when it is advisable to increase the rating of the existing fan. In one case, I replaced a 56 cubic-feet-per-minute (cfm) fan with a 115 cfm fan and cured a substantial problem with a large piece of medical equipment.

Keep in mind that heat rises. Therefore, fans and blowers that push air into the cabinet should be placed in the lower parts of the cabinet and the vent holes that exhaust the heated air on top. Similarly, if you prefer to use exhaust fans, then place the exhaust fan on top and the vent holes below.

DC Power Supplies for Digital and Other Circuits

The DC power supply is critical to the success of any electronic circuit, especially instrumentation circuits. This section deals with the control and distribution of DC power in digital and other instruments or equipment. We will look at the standard DC power supply circuit only in passing for those who may not be familiar with it.

DC Power Supplies

The purpose of the DC power supply is to convert alternating current (AC) to direct current (DC) for use in electronic circuits. *Figure 15-2* shows the basic circuit diagram for the DC power supply. Transformer T1 scales the voltage from 120 or 240VAC to the voltage range needed for the electronic circuits (typically 5 to 28VDC).

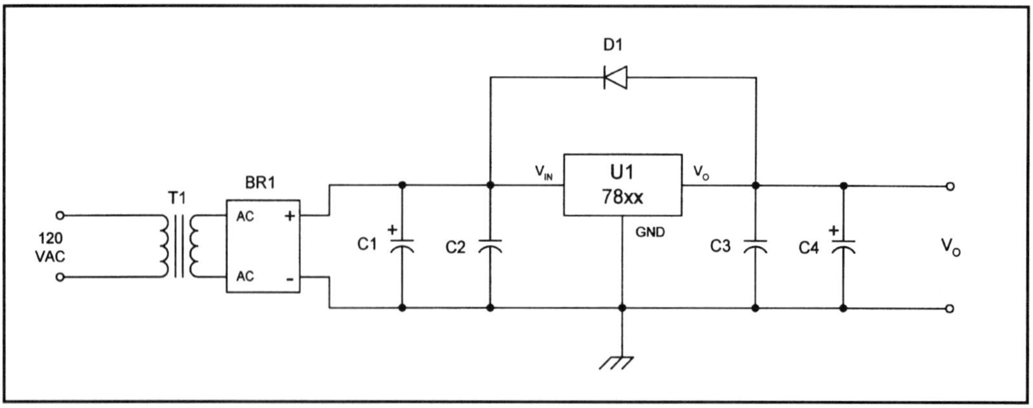

***Figure 15-2**. Typical regulated DC power supply circuit.*

The AC from the transformer is converted to DC in the bridge rectifier BR1. These four-terminal devices take the AC voltage at the terminals marked "AC" and outputs them through the DC terminals ("+" and "-"). The output of the rectifier is pulsating DC with a peak voltage about 0.9 times the peak AC voltage. The voltage rating for the transformer is in *rms* voltage, so the output of the rectifier is 0.9 x 1.414 x V_{RMS}, or 1.27V_{RMS}, where V_{RMS} is the voltage appearing across the secondary winding of the transformer.

The voltage regulator (U1) keeps the output voltage constant despite variations of input voltage or variations of load current. There is usually a 2.5V to 3V drop between the V_{IN} and V_O terminals of the voltage regulator. Thus, the transformer secondary rms voltage must be $\geq (V_O + 3)/1.27$.

The voltage regulator is usually a three-terminal IC device of either LM-340n-xx or 78xx types (essentially the same thing under different nomenclatures). In LM-340 devices, the "n" is replaced with the package style indicator: L for TO-92, K for TO-3, and T for TO-220. The "xx" is the voltage rating. Thus, an LM-340T-5 is a +5V regulator in a plastic TO-220 package. For negative output regulators, the LM-320n-xx series are used. In 78xx devices, the "xx" is the voltage rating; so a +5V type is 7805, while a +12 volt is 7812. If the device is rated at 100 mA or so, then an "L" is placed between the 78 and xx: 78L05 is a 100 mA, +5VDC device. Negative output equivalents are designated 79xx.

The current ratings for most regulators are:

TO-92	100 mA
TO-220	750 mA (1,000 mA with heat sink)
TO-3	1,000 mA (some are 3,000 mA)

Most common devices are either TO-92 (such as the 78L05) or TO-220 (such as the 7805), producing 100 mA and 750 mA, respectively.

Capacitor C1 in *Figure 15-2* is rated at 2,000 µF per ampere, except for low current devices in which case 1,000 µF is used; for the standard 750-1000 mA supply a 2,000 µF unit is used. The exact rating is not important, so long as the right range is achieved. The output capacitor (C4) is optional, and is used to smooth variations occurring before the regulator can catch up, or transients. This capacitor is rated nominally at 100 μF per ampere. In both C1 and C4, the voltage rating of the capacitor should be at least 150 percent of the applied voltage rating ... more if possible.

Capacitors C2 and C3 are used for noise suppression. These capacitors are needed to take care of transients coming across the AC power supply line. Make these capacitors 0.1 μF to 1 μF.

Diode D1 is used only if C4 is used. This diode should be a 1N4007 device. It is used to rapidly discharge C4 when the power is turned off, preventing it from biasing the substrate of U1 and thereby causing its destruction.

DC Power Distribution

The direct current (DC) power supply is one of the most important, but least considered, aspects of electronic project design. The elements of DC power supply design are well known, and have been covered in depth in these pages. But what is often missing is how to distribute the power throughout a project. And this little matter is of critical importance. When I was in engineering grad school, I worked for the university repairing medical equipment. A fairly large number of grad students and senior undergrads doing various projects found their way to my workshop with one problem or another. Very frequently, the design of the circuit was fine, but the implementation was a real problem. In most cases, they created their own bad situation by paying too little attention to the power supply distribution. The main errors were conductors that were too small and a lack of decoupling capacitors. One fellow

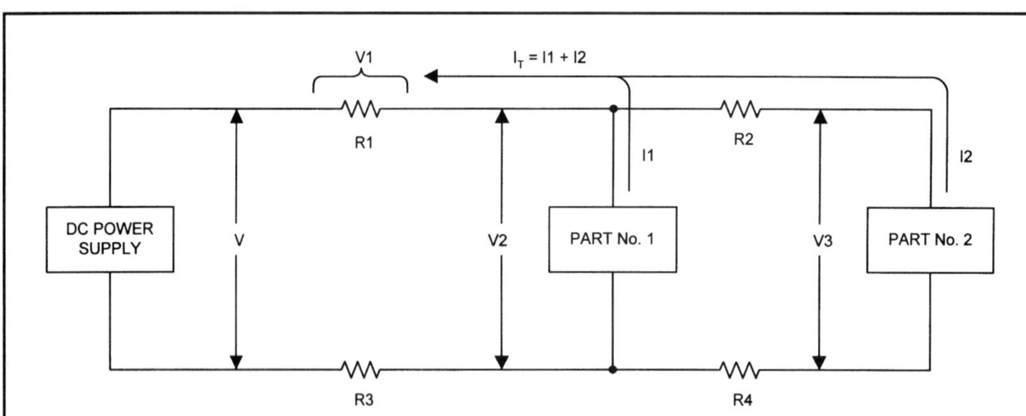

***Figure 15-3**. DC power distribution system.*

came into my shop with a perfboard covered with about thirty TTL chips, and complained that it worked only some of the time. After ascertaining that the problem was probably not a loose soldering connection or broken wire, I noticed that the wire to each chip was very small (#30 wire-wrap wire) and that there were no bypass capacitors anywhere on the perfboard. I instructed him to put 470 μF at the +5VDC input line, plus 0.1 μF at each TTL chip ... and increase the gauge of the wire distributing power. When he made the changes, the project worked properly.

In the later sections, you will find out why these things are important.

Figure 15-3 shows a generic electronic circuit DC power distribution system. A DC power supply, which may or may not be voltage regulated depending on the application, feeds DC to the individual circuits via a common DC power line and a ground. In an ideal world, this would be the end of our concerns. But in real circuits, there is a DC resistance associated with all of the power line conductors and any connectors used in the system. These resistances are shown as R1 through R4 in *Figure 15-3*. As current drawn by each part flows through the conductors, a voltage drop will appear across the conductor resistance. For example, voltage V1 is the IR drop across R1 due to the total current in the circuit.

Two things happen because of the voltage drops. First, the actual voltage available to each part is lower than the supply voltage by the amount of the voltage drops. For example, in Part 1, you can see a maximum voltage (V2) of V-V1. When selecting a value for V, you must either take into consideration the voltage drops that will occur or reduce the resistance to a negligible value by making the conductors large enough.

The second thing that can happen is that varying currents in one part will cause the supply voltage to vary at that part and at other parts. For example, if the DC load represented by current I1 varies with the signal in Part 1, then the total current varies, which means that the voltage drop across R1 also varies. The varying DC supply voltage is applied to Part 2 as its supply voltage. The signal can thus be coupled unintentionally from one stage to another through the DC power supply lines. This effect is usually called *power supply sensitivity* (PSS), and certain components (such as operational amplifiers) have a PSS specification.

The power supply DC return lines, which are usually also the grounds for the circuit, have the same kind of resistances as the "hot" line. If the same conductors are also used as signal grounds (often the case), then variations in ground voltage drops are seen as valid signals between stages; these are called *ground loops*. The equivalent circuit would show the varying DC from the ground loop in series with the normal signal applied to the stage. In a multistage circuit, the combination of voltage drops and spurious signal paths can be quite complex ... and very difficult to troubleshoot.

Fortunately, there are some things that you can do about it. One thing, of course, is to be generous with the wire size used in the circuit for DC power lines and grounds. I've seen problems caused by using #24 hookup wire where #18 would have been more appropriate. Also, I've seen cases where printed wiring tracks were too narrow and caused excessive voltage drops in relatively short lengths. My first homebrew microcomputer had this problem, because it contained a large number of old-fashioned TTL devices, each drawing 25 mA. The current requirement was huge.

Another thing that you can do about the problem is to be generous with the number of capacitors used in the circuit. The capacitors (like C1 and C2 in *Figure 15-4*) serve two functions: they decouple noise signals on the power line (prevent them from being coupled from stage to stage), and they serve as local reservoirs of current to fill in for short-duration current needs of the circuit being served. It has long been the practice in digital circuits to put a 0.1 μF capacitor across the power terminals of

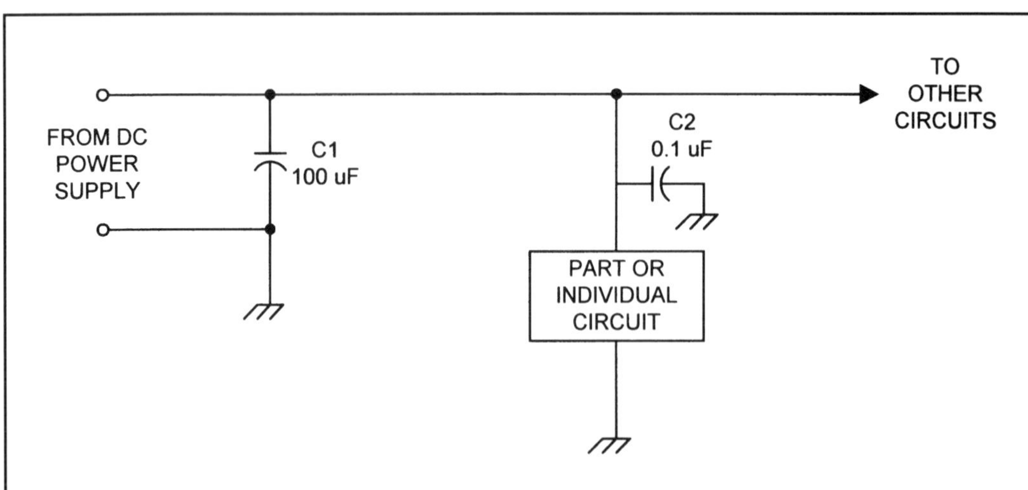

Figure 15-4. *Capacitor decoupling of power system.*

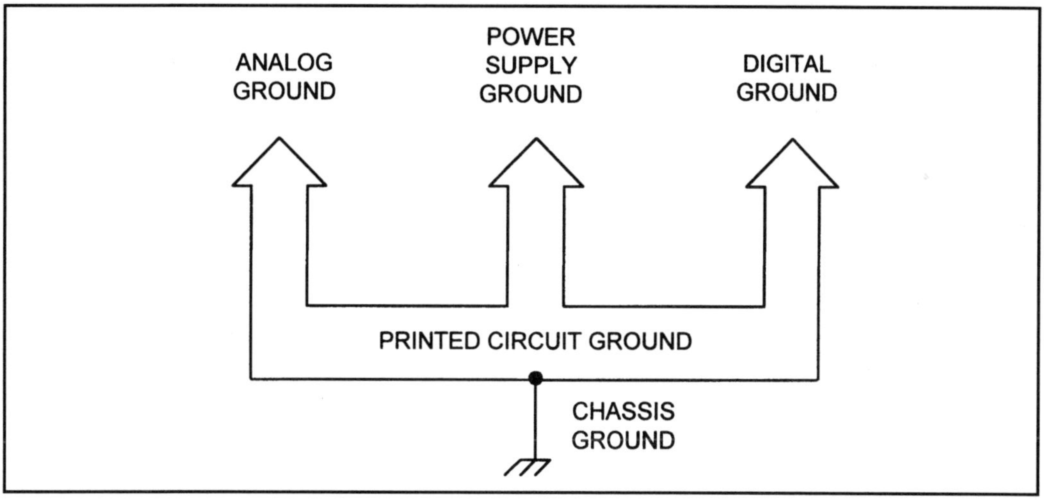

***Figure 15-5**. Use of multiple ground tracks to common point.*

each chip on the circuit board. Putting a 1 μF or higher capacitor at each analog circuit also works well.

Another thing that can be done is to use "single-point grounding" techniques. Each circuit should have its own ground system, with all of the individual ground systems connected to the chassis at only one point. Also, if you mix different kinds of grounds in the same project, then each type should have its own ground bus. In the example shown in *Figure 15-5*, there are separate grounds for DC power, analog signals and digital signals.

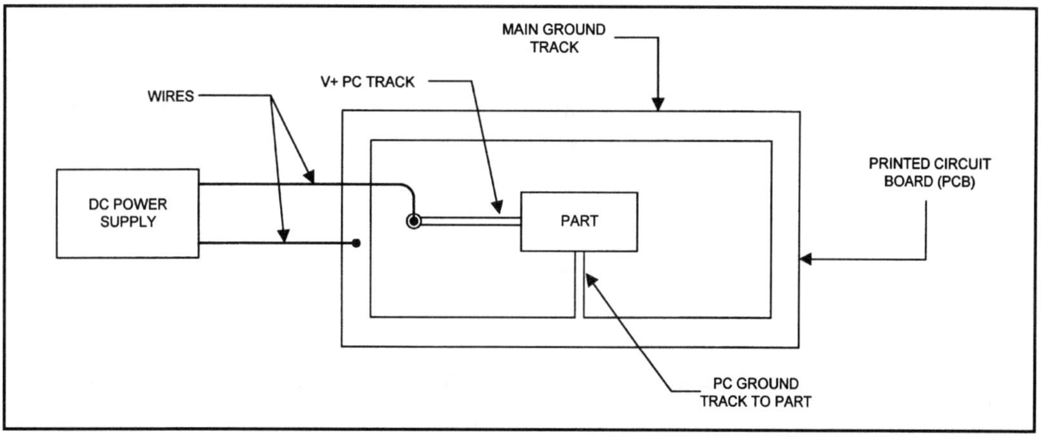

***Figure 15-6**. Printed circuit power distribution.*

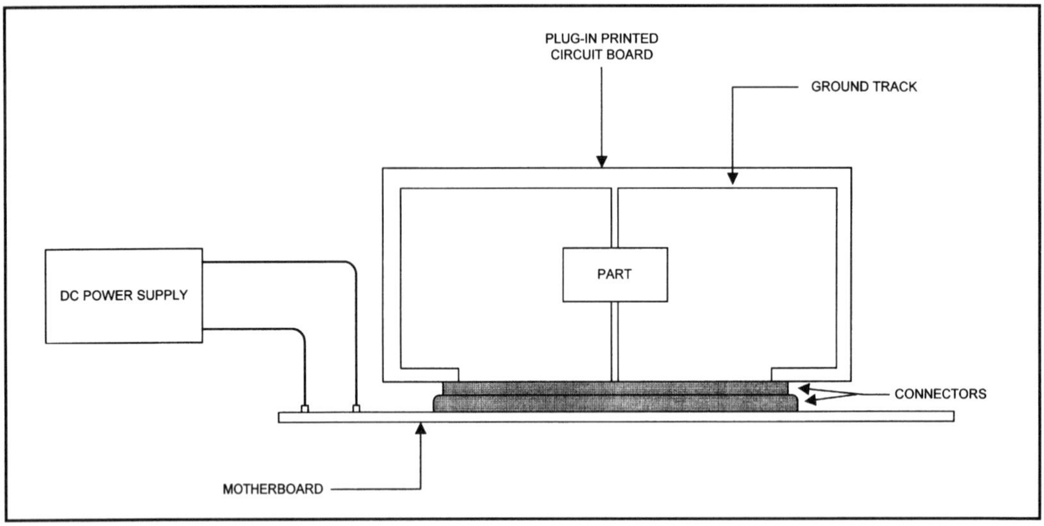

***Figure 15-7**. Power to a mother board.*

Power Distribution Schemes for Circuit Boards

There are a number of different approaches to managing DC power on projects with printed circuit boards. The version shown in *Figure 15-6* is the simple case of an individual circuit card served by an off-board DC power supply. Although in this case the negative polarity is grounded and the positive polarity is the "hot" line (the most common case), the opposite could be true as well. The common line is connected to the PCB ground tracks, while the hot line is connected to the V+ line. The DC is distributed to the individual parts or subcircuits through printed circuit tracks.

A different, but related scheme is shown in *Figure 15-7*. Here, there may be more than one individual printed circuit board with individual circuits, and all are plugged into a *motherboard*. The motherboard may be solely for the distribution of signals and DC power (as in the old S-100 microcomputers of the late 1970s and early 1980s), or there may be some additional circuitry on the motherboard. On the IBM-PC compatible computers, for example, the motherboard contains the processor and main functions of the board, and the plug-in PCBs perform I/O, video and other functions. In either case, the DC power is distributed to the various printed circuit boards via the motherboard and the plug-in connectors that mate the motherboards to the plug-in boards.

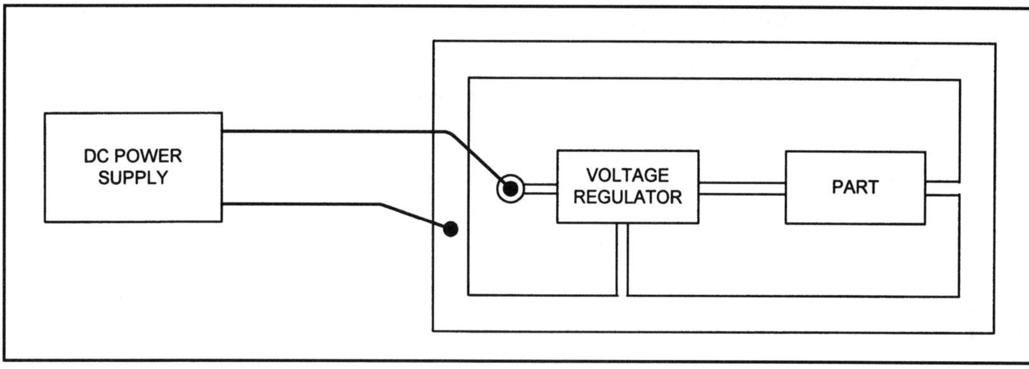

***Figure 15-8**. Distributed regulation scheme.*

The situation in *Figure 15-7* assumes that a single DC power supply serves the needs of all the PCBs plugged into the motherboard. In some cases, however, if current drains are high or there is a lot of digital (or other) noise generated on the PCBs, then one might wish to use an approach like *Figure 15-8*. This method is called *distributed voltage regulation* because each plug-in card has its own voltage regulator. The effect of the multiple regulators is to isolate the circuits on the board from ill effects caused by the operation of the other boards. The main DC power supply may be either regulated or unregulated. If a regulated main power supply is used, it is called *preregulation*, and it is capable of a very good degree of power supply ripple reduction.

The *power supply sense* method is shown in *Figure 15-9*. A DC power supply with sense lines samples the voltage present at a remote site, then adjusts its own voltage

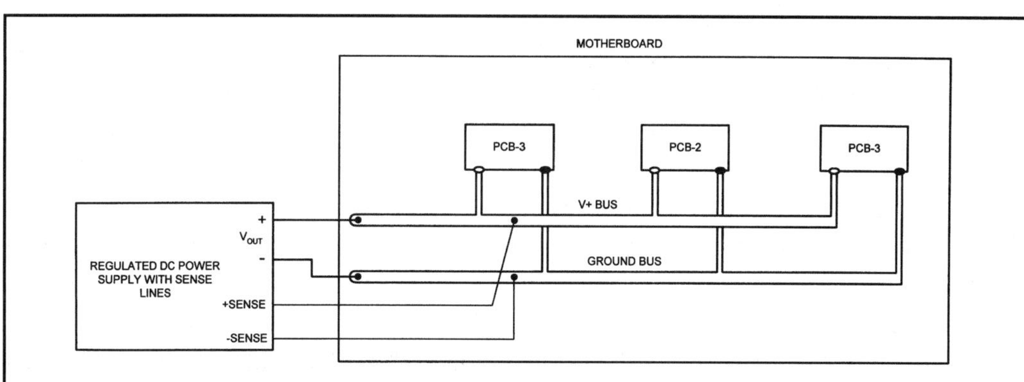

***Figure 15-9**. Sense regulation system.*

set-point to put the power supply output voltage high enough to meet the requirement at the remote site. For example, in a microcomputer that uses a motherboard and plug-in cards, it might be true that the processor card (PCB 2) is the most critical for voltage variation, and the memory cards are the least critical (PCB 1 or PCB 3). As a result, the builder will place the sense lines closest to the processor card, causing the DC power supply voltage to be +5VDC at this point. Closer to PCB 1, it will be higher, and toward PCB 3, it will be lower.

Isolating Circuits on the PCB

Figure 15-10 shows a way to improve the isolation of two or more circuits in a group. In most circuits, one would normally connect all the V+ lines to the positive DC power supply line, resulting in a potential feedback situation due to poor decoupling between stages. But the isolation between the two circuits can be improved if the V+ line to each circuit has a 10 to 100 ohm resistor (R1 and R2), and two additional decoupling bypass capacitors (C1 and C2). The main V+ line from the DC power supply is decoupled through a higher value capacitor (C3). The exact value of the resistances depends on the current drain, the V+ values required by each circuit, and the value of the supply voltage (VDC). At the normal current drains for each circuit, the voltage dropped across the isolation resistors should be only a small percentage of the total voltage. In radio frequency (RF) circuits, the resistors may be replaced with radio frequency chokes (RFCs).

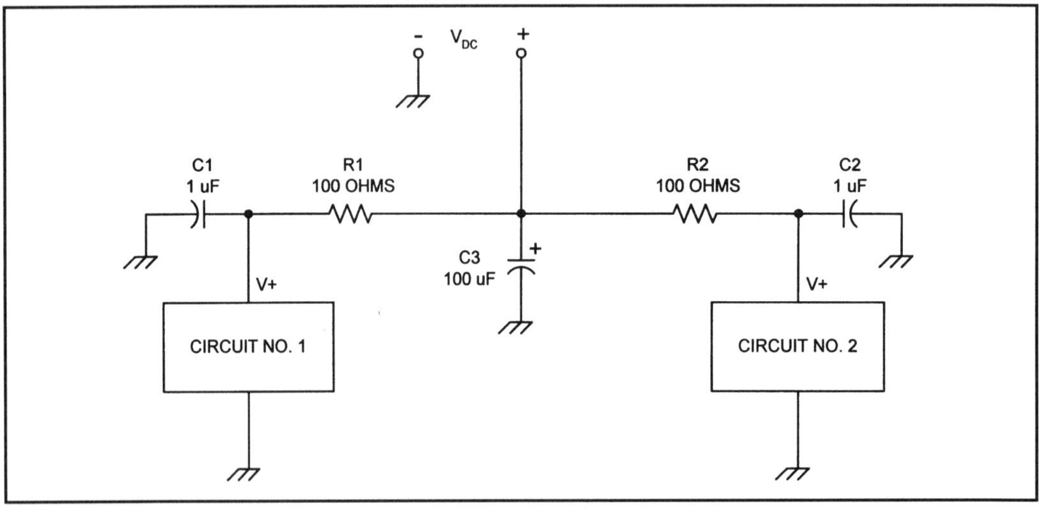

Figure 15-10*. R-C isolation system.*

Chapter 16

Using Microcontrollers in the Instrument Front End

A relatively new product has the potential to greatly improve the analog subsystem for sensor-based instruments, especially those that use remote sensors for a digital computer. Microcontrollers are small computers based on PIC technology. They consist of the PIC device and supporting circuitry on a small printed circuit board that can be connected into circuits like a component. Two basic forms of device were reviewed when writing this chapter: the Parallax Inc. *BASIC Stamp* products and the MicroMint *PicStic* products.

BASIC Stamp Products

Figure 16-1 shows a package outline for the Parallax BASIC Stamp model BS1-IC and the MicroMint PicStic 1, PicStic 2 and PicStic 3 devices. The 1.4 in. x 0.6 in. printed circuit board of the microcontroller mounts vertically on the printed circuit board with the other components. The connections are compatible with the 14-pin SIP socket. The more sophisticated Parallax BS2-IC (*Figure 16-2*) is a printed circuit, but is designed to mount in the socket of a 24-pin dual in-line package (DIP) integrated circuit.

The BS1-IC has eight bidirectional I/O lines, and executes up to 2,000 instructions per second. It will respond to up to 80 instructions. It provides a 2,400 baud serial I/

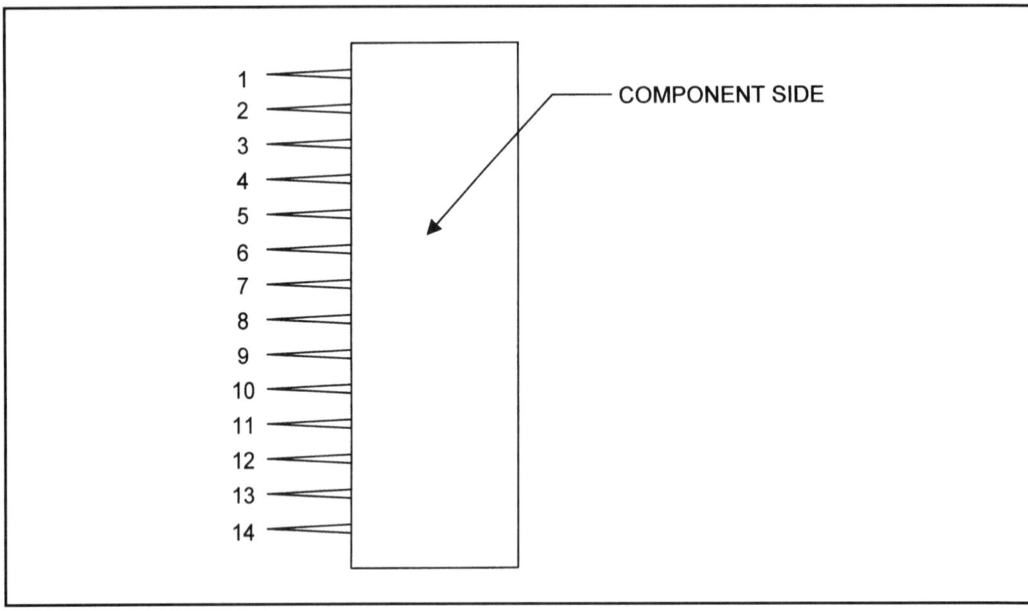

***Figure 16-1**. BASIC Stamp and PicStic device.*

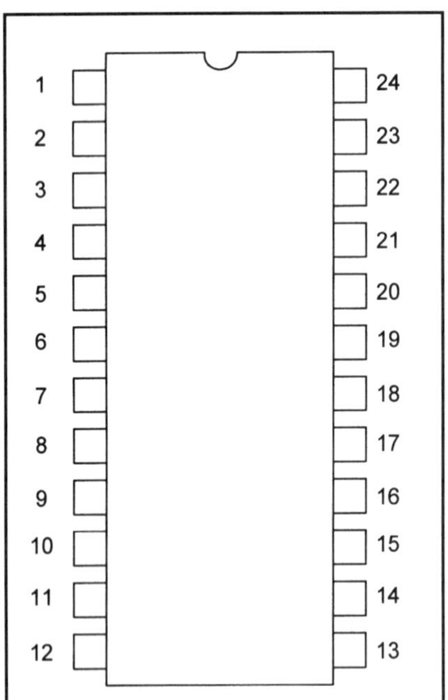

***Figure 16-2**. BASIC Stamp BS-2C device.*

O for connection to a larger computer or other instrument. Like the other products of this class, the BS1-IC will provide pulse measurements, and will recognize potentiometer settings (and hence resistance sensor values), pulse width modulation, sound and other features.

The BS2-IC is a larger version of the same concept and comes in a 24-pin DIP board. It provides 16 I/O lines, and allows up to 500 instructions which execute at a rate of 4,000 per second. The serial I/O operates up to 50K baud. The BS2-IC has the same features as the BS1-IC, but also offers touch-tones, frequency generation, pulse counters, serial shift registers, X-10 standard power line controls, and other features.

PicStic Products

There are three different PicStic products: PicStic 1, PicStic 2 and PicStic 3. In size and shape, they are similar to the BS1-IC device, but add some additional features. They have 1024 bytes of EEPROM memory. The PicStic devices use the same eight I/O lines as the BS1-IC, but also add two additional I/O pins and four interrupt sources (assembly language addressable only). The serial I/O operates at rates from 300 to 9,600 baud.

PicStic 1 is very similar to the BS1-IC, except for the additional pair of I/O lines. The PicStic 2 has an on-board real-time clock, while the PicStic 3 has an on-board, two-channel, 0-to-5V, 12-bit A/D converter.

Pinouts are:

	BS1-IC	**PicStic-1**	**PicStic-2**	**PicStic-3**
1	Power	Power	Power	Power
2	Ground	Ground	Ground	Ground
3	PCO	PA3	PA3*	PA3
4	PCI	PA4	PA4	PA4
5	+5VDC	+5VDC	+5VDC	+5VDC
6	NOT-RES	NOT-RES	NOT-RES	NOT-RES
7	PO	PBO	PBO	PBO
8	P1	PB1	PB1	PB1
9	P2	PB2	PB2	PB2
10	P3	PB3	PB3	PB3
11	P4	PB4	PB4	PB4
12	P5	PB5	PB5	PB5
13	P6	PB6	PB6	PB6
14	P7	PB7	PB7	PB7

— where:

POWER is DC power supply 6 to 15VDC.
GROUND is signal and power common.
+5VDC is a voltage output when POWER is used, or a connection to a +5VDC when POWER is not used.

PCO (BS1-IC) is PC-OUT to computer.
PCI (BS1-IC) is PC-IN from computer.
NOT-RES is an active-LOW reset.
P0-P7 (BS1-IC), ***PO-P15*** on BS2-IC, and ***PB0-PB7*** (PicStic) are individually configurable I/O pins.
PA3/PA4 are additional individually configurable I/O pins used on PicStic only.

The PicStic 2 requires an on-board 3V battery to keep the real-time clock chip active during power-off periods. The PicStic 3 has a pair of A/D converter analog inputs (***ADCin0*** and ***ADCin1***) that are on the board, rather than on a pinout.

The BS2-IC pinouts are:

1	Serial out	13	P8
2	Serial in	14	P9
3	RESET (HIGH)	15	P10
4	Serial Ground	16	P11
5	PO	17	P12
6	P1	18	P13
7	P2	19	P14
8	P3	20	P15
9	P4	21	+5VDC
10	P5	22	NOT-RES
11	P6	23	System ground
12	P7	24	POWER

Connecting to a Serial Port

All of the microcontrollers discussed so far can be connected to the serial I/O port on a personal computer. The standard RS-232C port found on these machines normally requires ±5V to ±12V signals for HIGH and LOW, but modern machines often work quite well with standard 0 (LOW) and +2.4 to +5V (HIGH) signals. In machines that will recognize TTL levels on the serial port, a circuit such as that shown in *Figure 16-3* can be used. In this circuit, the serial-out pin of the microcontroller is connected to pin 2 of the DB-9 nine-pin RS-232 connector commonly used on today's computers. The serial-in pin of the microcontroller is connected to pin number 3 of the DB-9 connector through a 22 kohm resistor.

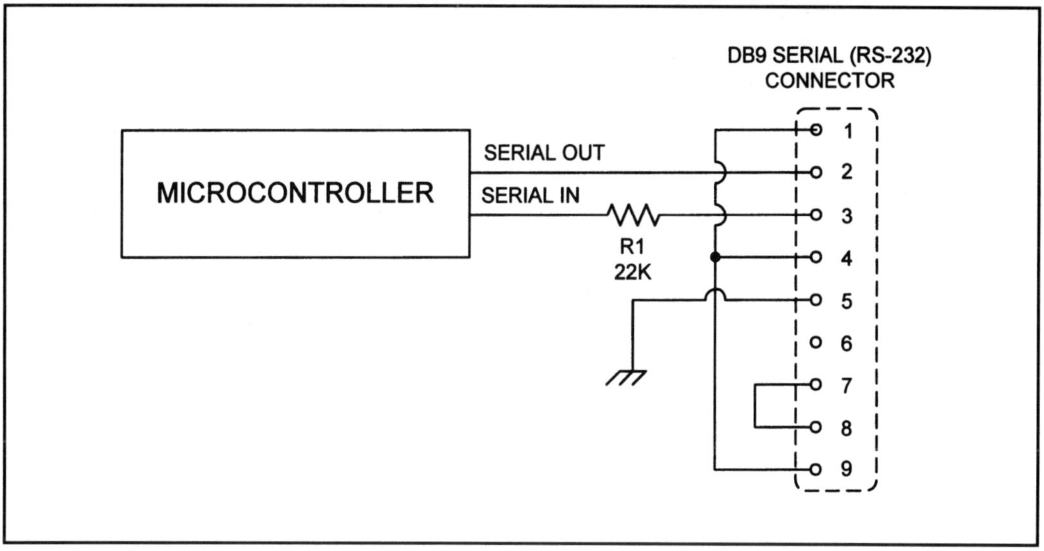

Figure 16-3. *Interfacing to RS-232C.*

For cases where the computer does not respond to TTL levels, or where you wish to use the standard signal levels, then use one of the RS-232C interface chips shown in *Figure 16-4* (LT1181ACN) and *Figure 16-5* (MAX-232). These chips operate from +5 volts, but produce the negative signal level using a capacitor charge-pump circuit internally.

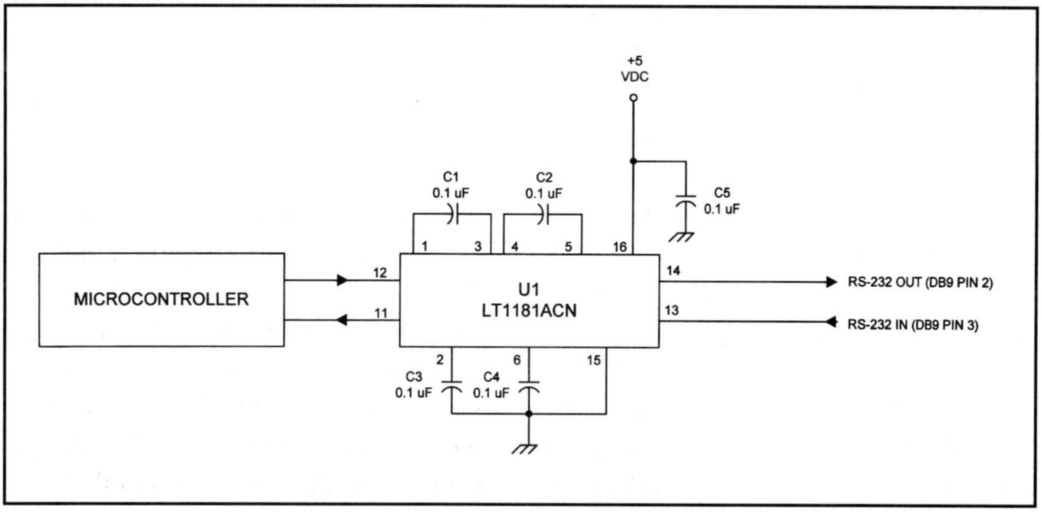

Figure 16-4. *Interfacing to RS-232C with interface chip.*

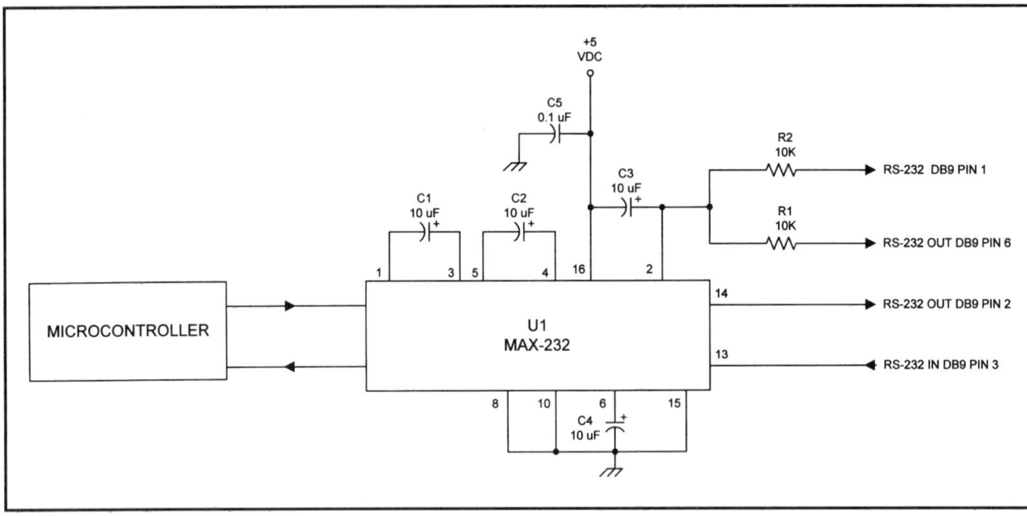

Figure 16-5. Use of MAX-232 interface.

Push-Button Interfacing

It is easy to interface a push-button or switch to the microcontrollers by using one of the I/O pins, as shown in *Figure 16-6*. The I/O pin is configured as an input. Resistor R1 is connected to the +5VDC power supply, and is used to force the input HIGH whenever S1 is open. If S1 is closed, the input pin is grounded, so is seen as LOW by the computer. The on/off state of switch S1 is read by programming the

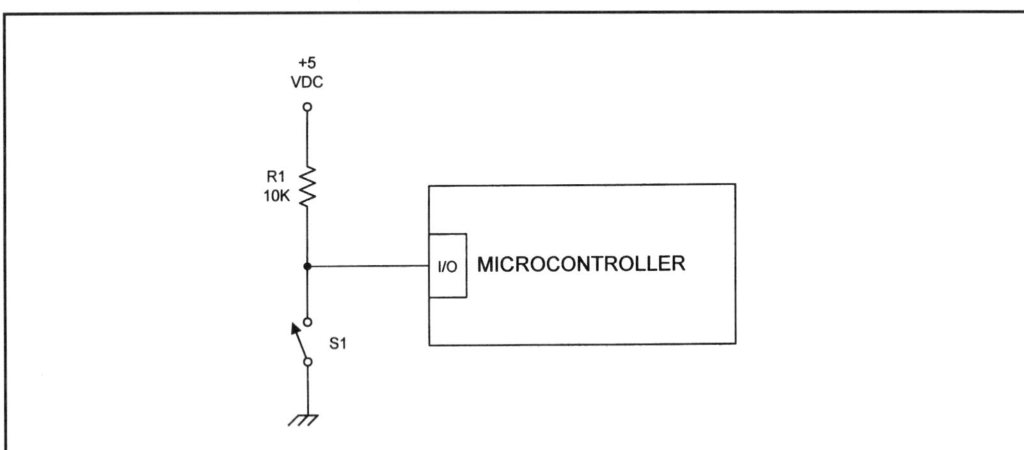

Figure 16-6. Adding a switch to a pin.

microcontroller. Some circuits omit R1, but that is not a good practice because noise can temporarily assert a LOW on the otherwise HIGH pin. The purpose of R1 is to force the pin to be HIGH when S1 is open.

Potentiometer Connection

Figure 16-7 shows the connection of a potentiometer to any I/O pin of the microcontroller. The microcontroller reads the value of the potentiometer (5 to 50 kohms) by measuring how long it takes to discharge the 0.1 μF capacitor through the I/O pin. The dialect of integer BASIC used by the microcontrollers has a POT command that reads this value.

The potentiometer could be any resistive sensor, such as an actual potentiometer, a photocell, a thermistor or a strain gauge.

A/D Converter

For those microcontrollers that do not have a built-in analog-to-digital converter (A/D), the circuit of *Figure 16-8* can be used. This circuit uses one of several 12-bit serial output A/D converters that are on the market (LTC1298 is used here, but MAX-

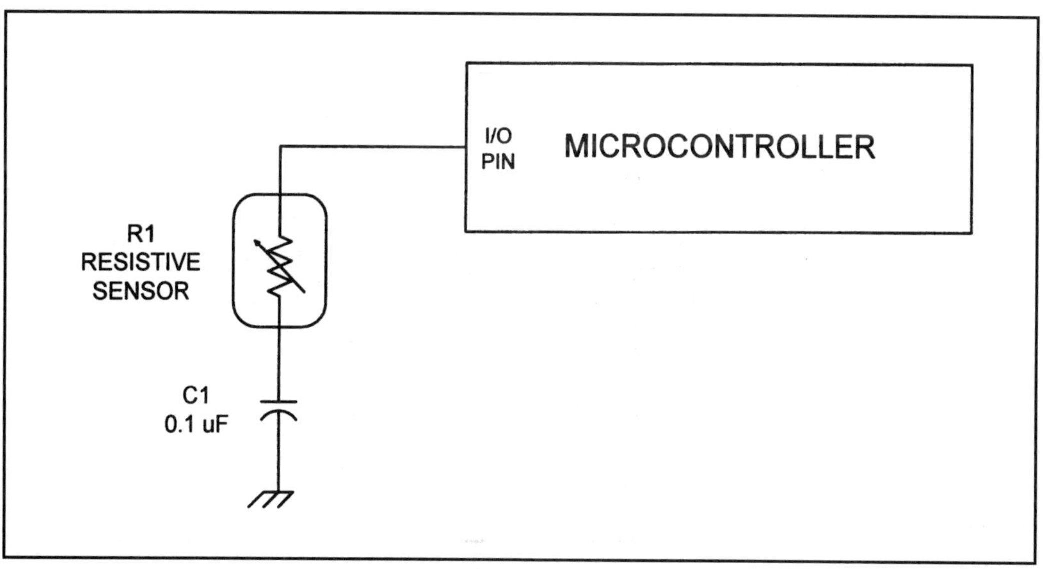

Figure 16-7. *Resistive sensor to a pin.*

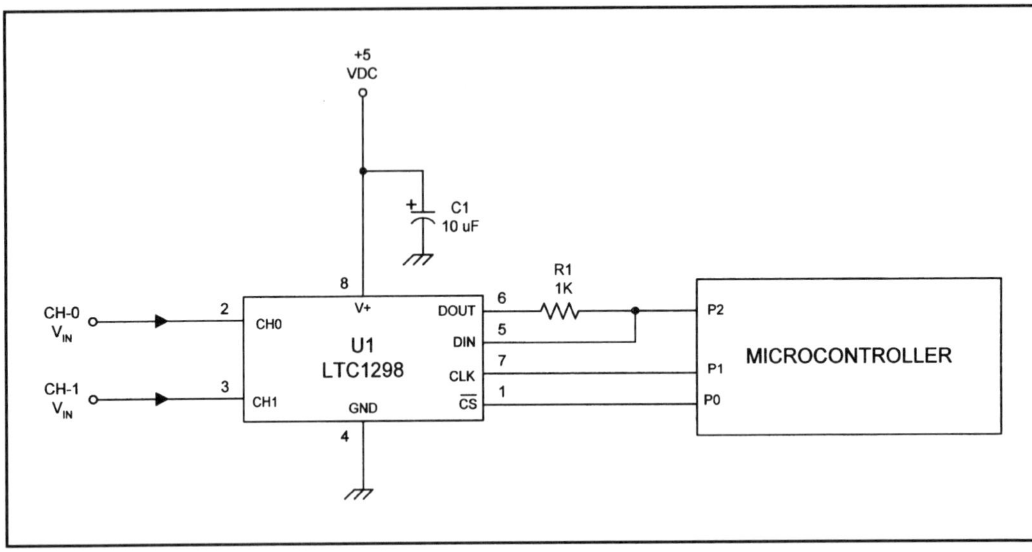

Figure 16-8 A/D controller.

186 and MAX-187 can also be used with different pinouts). This particular circuit has two 12-bit, 0-5V A/D channels that can be selected using the software instructions programmed into the microcontroller. The Parallax manual for the BS1-IC and BS2-IC devices also shows the circuit for an ADC-0831 serial A/D converter.

Programming

The microcontrollers can be programmed in C, assembly language, or a special version of integer BASIC called PBASIC. The manufacturers sell carrier boards for interfacing the computer (for programming the microcontroller) and other components. They also sell programming and development kits that assist the circuit developer in making the microcontrollers work in practical circuits.

Other Sources

A number of Stamp-related products can be obtained from Scott Edwards Electronics. I have used the A/D data acquisition board. Although it looks a bit pricey, it is an instrument in its own right and worth the price. Also, see Scott Edwards' column in *Nuts 'n' Volts* magazine for practical projects using Basic Stamps.

Connections

Parallax Inc.
3805 Atherton Road, #102
Rocklin, CA 95765
(916) 624-8333 (voice)
(916) 624-8003 (FAX)
Web: http://www.parallaxinc.com
ftp: ftp.parallaxinc.com
E-mail: info@parallaxinc.com
List server: majordomo@parallaxinc.com

MicroMint, Inc.
4-Park Street
Vernon, CT 06066
1-800-635-3355 (voice/orders)
860-871-6170 (voice)
860-872-2204 (FAX)
860-871-1988 (BBS)
Web: http://www.micromint.com

Scott Edwards Electronics
P.O. Box 160
Sierra Vista, AZ 85636-0160
520-459-4802 (voice)
520-459-0623 (FAX)
E-mail: 72037.2612@compuserve.com

Chapter 17

Data Converters (A/D and D/A)

Data converters do one of two jobs:

1. Convert a binary digital word to an equivalent analog current or voltage.
2. Convert an analog current or voltage to an equivalent binary word.

The former are called *digital-to-analog converters* (DAC), while the latter are called *analog-to-digital converters* (A/D or ADC). These devices form the interface between digital computers and the analog world ... which includes most sensors and actuators.

Approaches to Data Conversion

There are several approaches to converting data from electronic circuits and scientific instruments into digital binary numbers required by a digital computer. This job requires an A/D converter. The opposite type of converter (DAC) does the opposite work. It converts the binary output from a computer into an analog voltage or current. In discussing data converters, it is best to begin with a discussion of DACs because many A/D circuits contain embedded DACs.

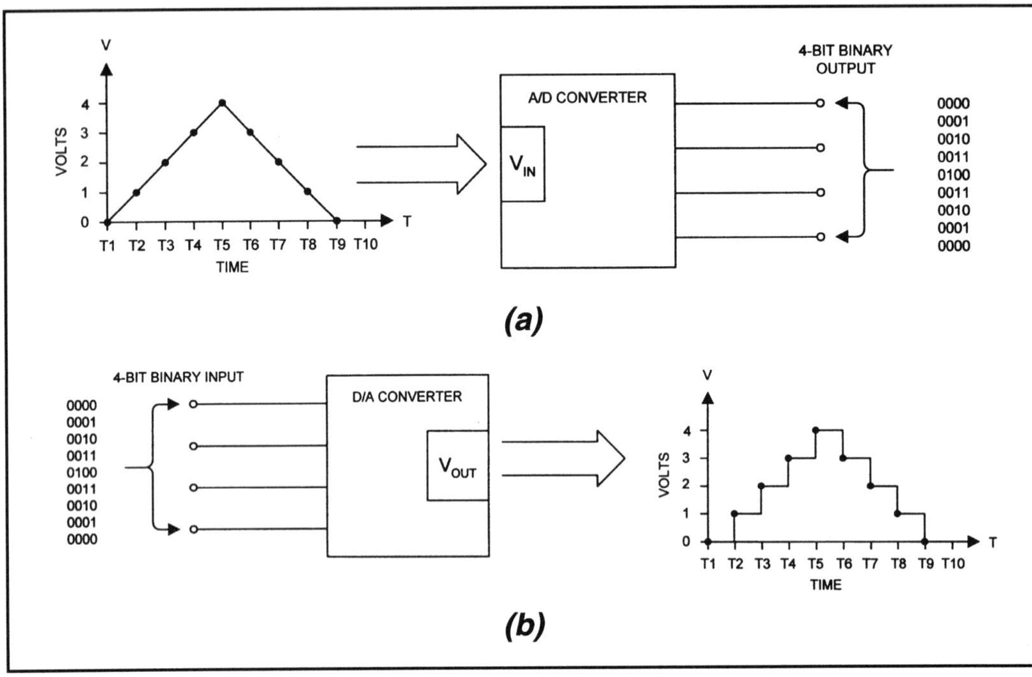

Figure 17-1*. a. The analog-to-digital conversion function; b. the digital-to-analog conversion function.*

The differences between A/D and DAC functions are shown in *Figure 17-1a* and *Figure 17-1b*. In *Figure 17-1a*, an A/D converter is used to translate an analog voltage waveform into the equivalent binary numbers that the computer can use. The waveform is simplified compared with actual waveforms, but this is done to make it easier for me to write and you to read. This achieves certain values of voltage between 0 and 4 volts, as follows:

Time	**Voltage**
T1	0 volts
T2	1 volt
T3	2 volts
T4	3 volts
T5	4 volt
T6	3 volts
T7	2 volts
T8	1 volt
T9	0 volts

This voltage waveform is applied to the analog input of the A/D converter, which produces the following output binary numbers at the times indicated:

Time	Binary Output	Decimal Equivalent
T1	0000_2	0 volts
T2	0001_2	1 volt
T3	0010_2	2 volts
T4	0011_2	3 volts
T5	0100_2	4 volts
T6	0011_2	3 volts
T7	0010_2	2 volts
T8	0001_2	1 volts
T9	0000_2	0 volts

In other words, the A/D converter takes the analog input voltage, converts it into a series of binary (base 2) numbers ("words") that represent the values, then forwards these binary numbers to an input port of the computer.

The DAC process is just the opposite (*Figure 17-1b*). In this case, the digital computer generates a series of binary values that represent a data pattern from which the waveform can be constructed. The DAC will convert the binary numbers into equivalent voltages (or currents in many models), and output them to instruments such as voltmeters, ammeters, oscilloscopes or mechanical (paper) chart recorders.

Note the difference in waveforms between *Figure 17-1a* and *Figure 17-1b*. They are supposedly the same because the same set of binary numbers is associated with each. The waveform in *Figure 17-1b* is what you get if you digitize the waveform in *Figure 17-1a* then feed it to a DAC for the reconstruction of the analog waveform. The distortion is due to the fact that too few bits were used to represent the waveform. If more bits (such as 12 or 16 bits) were used, then the waveform in *Figure 17-1b* would more nearly resemble that of *Figure 17-1a*.

The A/D converter can be internal; that is, on a plug-in printed circuit board that is interfaced directly with the internal computer data bus structure. Alternatively, the A/D converter can be interfaced to the computer via one of the ports (*Figure 17-2*). There are three different forms of port: 8-bit or 16-bit parallel port, serial communications port and Centronics printer port. The straight parallel ports are rare, and

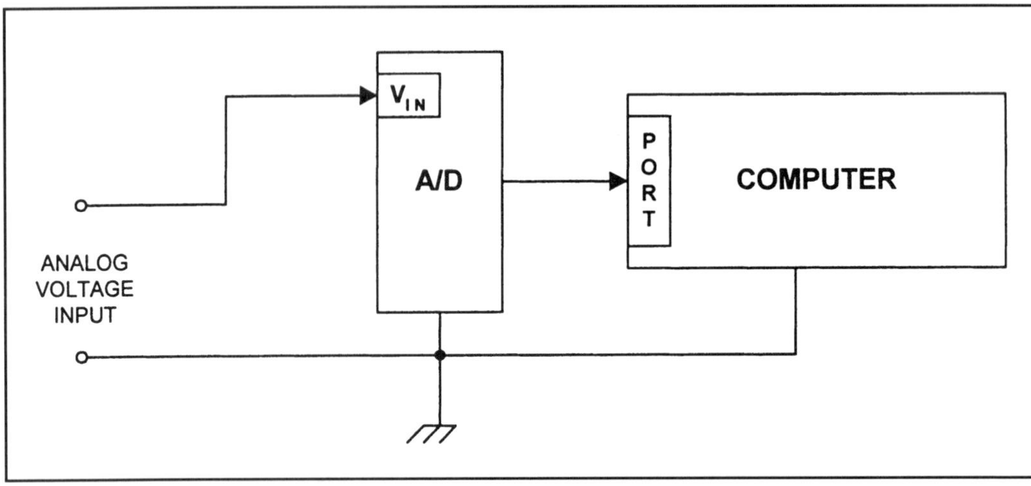

***Figure 17-2**. Interfacing the A/D converter to computer.*

must be purchased as separate plug-in cards for most PCs. These ports consist of 8 or 16 separate input lines for carrying data (one line for each data bit).

The serial communications port is the RS-232 port on the back of your computer. It's on most computers; up to four of them can be included with your PC, designated COM1, COM2, COM3 or COM4. The printer port is normally an output port, so it is not often thought of as an input port. But there are several handshaking lines on the printer port that are bidirectional. They can be used to trigger the START function of the A/D, or receive data or end-of-conversion signals.

Many of the A/D converters that we will examine are based on the DAC circuit, so (even though DACs *per se* are of less interest to most) we will look at DAC circuitry first.

Digital-to-Analog Converters (DACs)

There are several different approaches to DAC design, but all of them are varieties of a weighted current or voltage system that accepts binary words by appropriate switch contacts. The most common example is the *R-2R ladder* shown in *Figure 17-3*. The active element, A1, is an operational amplifier in a unity gain inverting follower configuration. In the circuit of *Figure 17-3*, the digital inputs are shown as mechanical switches; but in a real data converter circuit, the switches would be re-

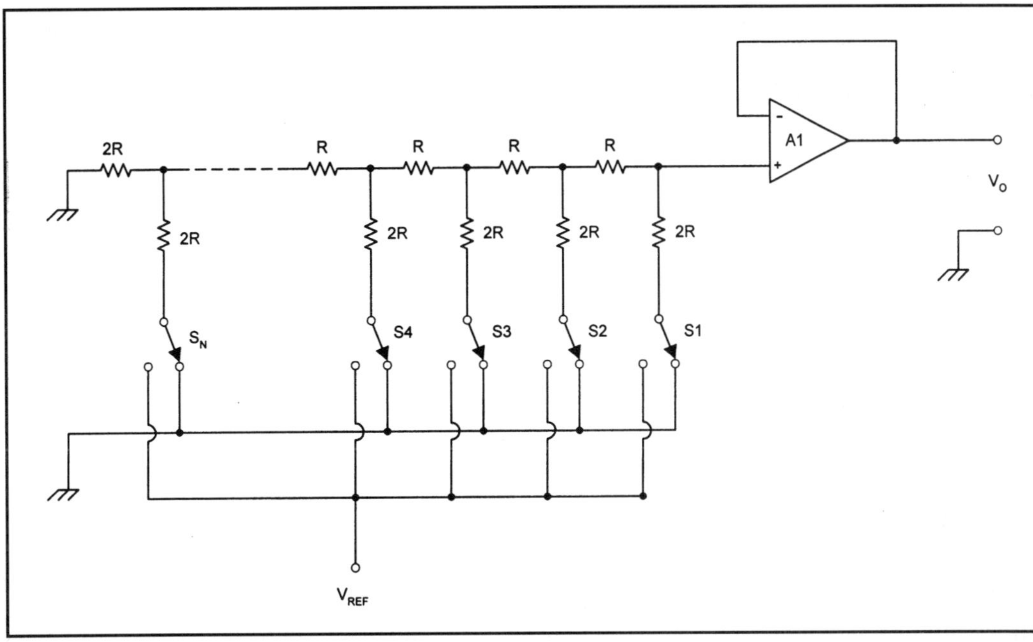

***Figure 17-3**. R-2R binary ladder.*

placed by electronic switching devices (such as transistors). The electronic switches are driven by either a binary counter or an N-bit parallel data line.

A *precision reference voltage* source (Vref) is required for accurate data conversion. This voltage is most often +2.56 volts, +5.00 volts or +10.00 volts. Other voltages can also be used, however. The accuracy of the converter is dependent upon the precision of the reference voltage source. There are other sources of error, but if the reference voltage accuracy is poor, then there is no hope for any other factors to be effective in improving the performance of the circuit. I can recall situations where expensive 16-bit data converters were used with zener diode reference sources that had so much thermal drift that they were suited only to 4 to 6-bit models. Although almost any voltage regulator can be pressed into service as the reference, it is prudent to select a precision, low-drift model. Consider the following situation: a 12-bit A/D converter breaks up the input voltage range into 4,096 different discrete values; or if one is zero, 4,095 different nonzero values. Each step, produced by a change in the least significant bit of the 12-bit binary word, represents 0.0024 volts, or 2.4 mV (this is called the "1-LSB" value). On a "0 to 10 volt" A/D converter, therefore, the maximum input voltage is 9.9976 volts (with some doubt in the last position). If a zener diode with a thermal drift of ±0.100 volts (100 mV) is used as the reference, then the uncertainty arises at the 9.90V level. This is a good way to make a 12-bit A/D converter act like a 6- or 7-bit A/D.

Returning to *Figure 17-3*, consider the circuit action under circumstances where various binary bits are either HIGH or LOW. If all bits are LOW, then the output voltage will be zero. The value of the output voltage is given by the product I x R, and when all bits are LOW, this current is zero. In practical circuits, though, there might be some output voltage under these circumstances due to offsets in the operational amplifier, the R-2R ladder and the electronic switches. These offsets can be nulled to zero output voltage when all bits are intentionally set to zero (or ignored, if negligible).

The unterminated R-2R ladder produces an output current. Some commercial IC DACs are current output models, and have no output amplifier. If there is a terminating resistor (R_t) shunting the output terminals of the DAC, then the circuit produces an output voltage I_o x R_t. The output impedance of such a circuit tends to be high, so some of these DACs use an output amplifier in order to produce a low-impedance voltage output. The transfer function of the R-2R ladder type of DAC is:

$$V_o = \frac{V_{ref} A}{2^N}$$

— where:

> V_o is the output potential.
> V_{ref} is the reference potential.
> A is the decimal value of the applied binary word.
> N is the number of bits in the applied binary word.

If the most significant bit (MSB) is made "1" (or HIGH), then the output voltage will be approximately $V_{ref}/2$. Similarly, if the next most significant bit is turned on (set to HIGH) and all others are LOW, then the output will be $V_{ref}/4$. The least significant bit (LSB) would contribute $V_{ref}/2N$ to the total output voltage. For example, with an 8-bit DAC, the LSB changes the output $V_{ref}/28$, or $V_{ref}/256$. This change is called the *1-LSB value* because it is the change that occurs in response to a change in the least significant bit. *Figure 17-4* graphs the output of a voltage DAC in response to the entire range of binary numbers applied to the digital inputs. The result is a *staircase*

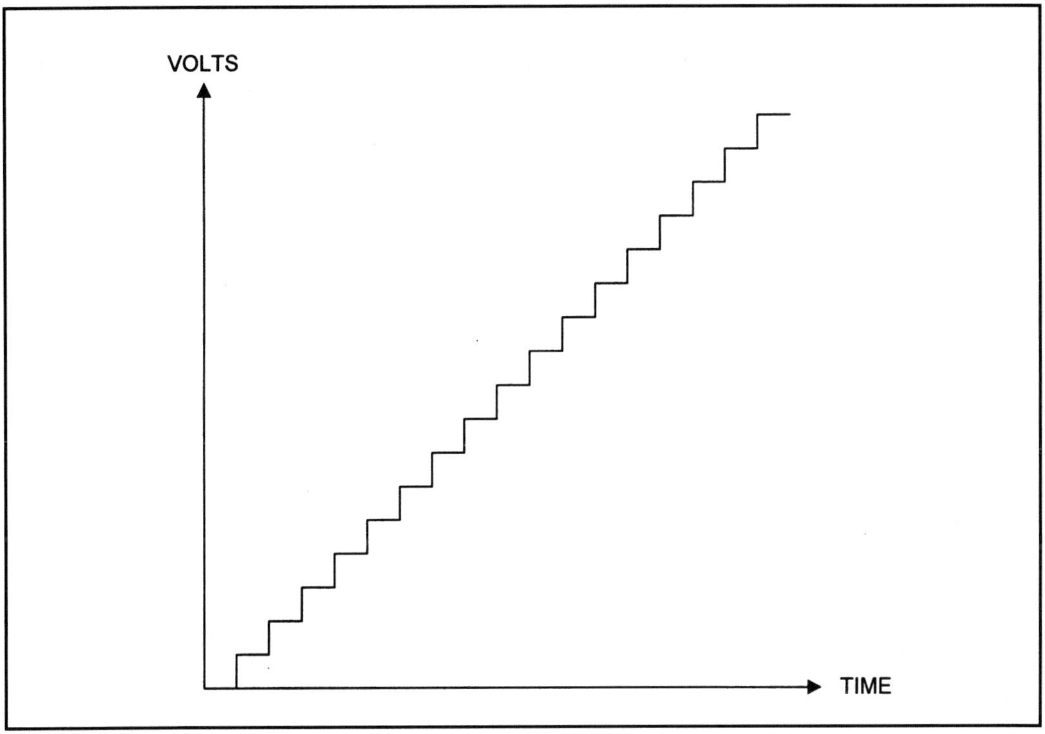

Figure 17-4*. Waveform of a DAC output when input runs through the binary numbers.*

waveform that rises by the 1-LSB value for each 1-LSB change of the binary word. This step height represents the minimum discernible resolution of the circuit.

The reference source can be either internal or external to an integrated circuit DAC. If the reference voltage or current source is external, then the DAC is said to be a *multiplying DAC* or MDAC. The multiplication takes place between the analog reference and the fraction defined as $A/2^N$ in the previous equation. If the reference source is completely internal, and not adjustable (except for fine trimming), then the DAC is said to be a *non-multiplying DAC* or simply *DAC*.

Coding Schemes

There are several different coding methods for defining the transfer function of the DAC. The most common of these are: *unipolar positive*, *unipolar negative*, *symmetrical bipolar* and *asymmetrical bipolar*.

The unipolar coding schemes provide an output voltage of one polarity only. These circuits usually produce 0 volts for the minimum and some positive or negative value for the maximum. Because one binary number state represents 0VDC, there are [2^N - 1] states to represent the analog voltages within the range. For example, in an 8-bit system there are 256 states, so if one state (00000000_2) represents 0VDC then there are 255 possible states for the nonzero voltages. Thus, the maximum output voltage is always 1-LSB less than the reference voltage. For example, in a 10.00VDC system, the maximum output voltage is 10.00 volts/256 = 0.039 volts = 39 mV less. Thus, the maximum output voltage will be approximately 9.96VDC. The unipolar positive coding scheme is:

Unipolar Output	**Voltage Binary Input Word**
0.00VDC	00000000_2
$V_{max}/2$	10000000_2
V_{max}	11111111_2

The negative version of this coding scheme (unipolar negative) is identical, except that the midscale voltage is $-V_{max}/2$ and the full-scale output voltage is $-V_{max}$. A variant on the theme inverts the definition so that:

Unipolar Output	**Voltage Binary Input Word**
0.00VDC	11111111_2
$V_{max}/$	1000000_2
V_{max}	$00000000_{@-\{2\}}$

The bipolar coding scheme faces a difficulty that requires a trade-off in the design. There is an even number of output states in a binary system. For example, in the standard 8-bit system, there are 256 different output states. If one state is selected to represent 0VDC, then there are 255 states left to represent the voltage range. As a result, there is an even number of states to represent positive and negative states either side of 0VDC. For example, 127 states might be assigned to represent negative voltages, and 128 to represent positive voltages. In the asymmetrical bipolar coding, therefore, the pattern might look like:

Bipolar Output Voltage	**Binary Input Word**
$-V_{max}$	00000000_2
0.00VDC	10000000_2
$+V_{max}$	11111111_2

A decision must be made regarding which polarity will lose a small amount of dynamic range.

The other bipolar coding system is the symmetrical bipolar scheme. The decision in the symmetrical scheme is that each polarity will be represented by the same number of binary states on either side of 0VDC. But this scheme does not permit a dedicated state for zero. The scheme is:

Bipolar Output Voltage	Binary Input Word
$-V_{max}$	00000000_2
-Zero (-1 LSB)	01111111_2
0.00VDC	(Disallowed)
+Zero (+1 LSB)	10000000_2
$+V_{max}$	11111111_2

The state "plus zero" is more positive than 0VDC the 1-LSB value, while the "minus zero" state is more negative than 0VDC by the same 1-LSB value.

A Practical IC DAC Example

A number of different manufacturers offer low-cost IC DACs that contain almost all of the circuitry needed for the process, except possibly the reference source (although some devices also contain the reference source) and some operational amplifiers for either level shifting or current-to-voltage conversion.

For the purposes of this chapter, the DAC-08 device is used as a practical circuit example. This eight-bit DAC is now something of an industry standard, and is available from several sources. This DAC is sometimes designated LMDAC-0800. An easily-available and closely-related device is the DAC-0806.

Figure 17-5a shows the basic circuit configuration for the DAC-08. In subsequent circuits, the power supply terminals are deleted for simplicity's sake; they will always be the same as shown here. The internal circuitry of the DAC-08 is the R-2R ladder shown in the previous section, but it has two outputs: I_o and NOT-I_o. These current outputs are unipolar and complementary (*Figure 17-5b*); if the full-scale output current is I_{max}, then $I_{max} = [I_o + \text{NOT-}I_o]$. The specified value of I_{max} on the DAC-08 family of DACs is 2 mA.

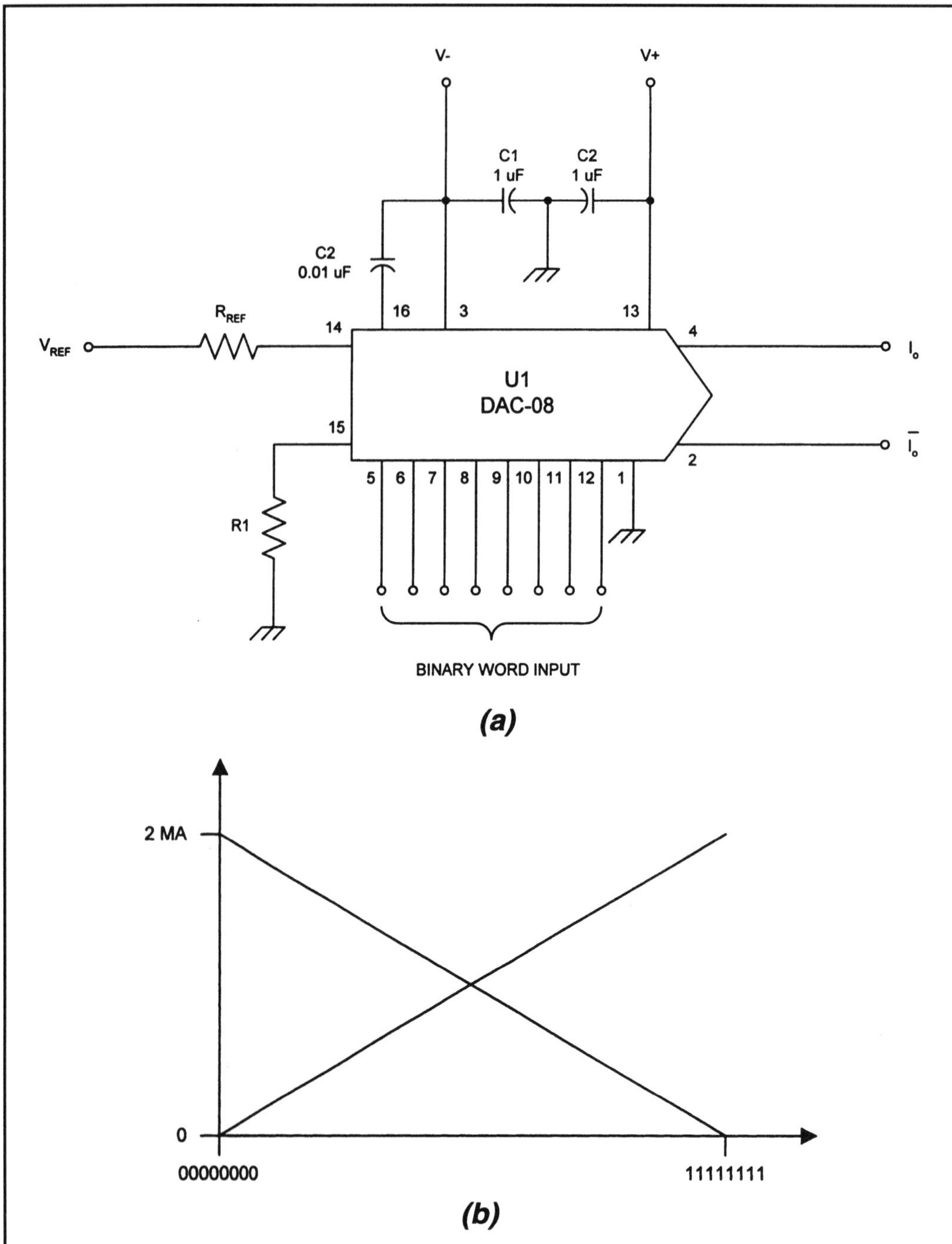

***Figure 17-5**. a. Basic DAC-08 circuit; b. output function.*

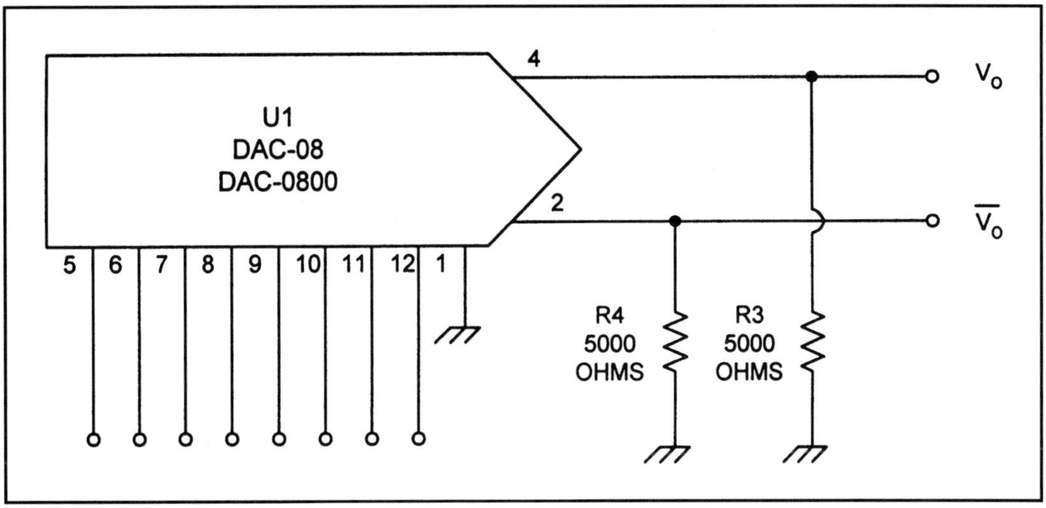

Figure 17-6*. 0-10 volt, TTL compatible circuit with voltage output.*

Two types of input signal are required to make this DAC work: an *analog reference* and an *eight-bit digital* signal. The analog signal in the reference current, I_{ref}, is applied through pin number 14. This current may be generated by combining a precision reference voltage source with a precision low-temperature coefficient resistor to convert V_{ref} to I_{ref}. Alternatively, a constant current source may be used to provide I_{ref}. For TTL compatibility of the binary inputs, make V_{ref} = 10.000 volts, and R_{ref} = 5000 ohms.

The other type of input is the eight-bit digital word, which is applied to the IC at pins 5 through 12, as shown. The logic levels which operate these inputs can be preset by the voltage applied to pin number 1 (for TTL operation, pin number 1 is grounded). In the TTL-compatible configuration shown, LOW is 0 to 0.8 volts, while HIGH is +2.4 to +5 volts.

Figure 17-6 shows the connection of the DAC-08 (less power supply and reference input) required to provide the simplest form of unipolar operation over the range of approximately 0 to 10 volts. When the input word is 000000002, then the DAC output is 0 volts, plus-or-minus the DC offset error. A half-scale voltage (-5 volts) is given when the input word is 10000000_2. This situation occurs when the MSB is HIGH and all other digital inputs are LOW. The full-scale output will exist only when the input word is 11111111_2 (all HIGH). The output under full-scale conditions will be -9.96 volts rather than 10 volts, as might be expected (note: 9.96 volts is 1-LSB less than 10 volts). These codes are:

Condition	**Binary Code**	V_o
Full scale	11111111	-9.96 volts
Half scale	10000000	-5.00 volts
Zero	00000000	0.0 volts

The circuit in *Figure 17-6* works by using resistors R2 and R3 as current-to-voltage converters. When currents I_o and NOT-I_o pass through these resistors, a voltage drop of IR, or 5.00 x I_o (mA) is created. A problem with this circuit is that it has a high source impedance (5 kohms, with the values shown for R2/R3).

Figure 17-7 shows a simple method for converting I_o to an output voltage (V_o) with a low output impedance (less than 100 ohms) by using an inverting follower operational amplifier. The output voltage is simply the product of the output current and the negative feedback resistor:

$$V_o = R \times I_o$$

As in the case previously described, a 5000 ohm resistor and a 10.00VDC reference voltage will produce a 9.96V output voltage when the DAC-08 is set up for TTL inputs and 2.0 mA $I_{o(max)}$.

The frequency response of the DAC circuit can be tailored to meet certain requirements. The normal output waveform of the DAC is a staircase when the digital input increments up from 00000000_2 to 11111111_2 in a monotonic manner. In order to make the staircase into an actual ramp function, a low-pass filter is needed at the output to remove the "stepness" of the normal waveform. A capacitor shunted across the feedback resistor (R) will offer limited filtering on the order of -6 dB/octave above a cutoff frequency of:

$$F = \frac{1{,}000{,}000}{2 \pi R C_{\mu F}}$$

— where:

F is the -3 dB frequency in hertz (Hz).
R is in ohms (Ω).
CmF is in microfarads (μF).

In most practical circuits, the required value of F is known from the application. It is the highest frequency Fourier component in the input waveform. It is necessary to calculate the value of capacitor needed to achieve the cut-off frequency, so we would swap the F and C terms:

$$C_{\mu F} = \frac{1{,}000{,}000}{2 \pi R F}$$

A/D Converters

The *analog-to-digital converter* (A/D) is used to convert an analog voltage or current input to an output binary word that can be used by a computer. Of the many techniques that have been published for performing an A/D conversion, only a few are of interest to us. So, we will consider only the *voltage-to-frequency*, *single-slope integrator*, *dual-slope integrator*, *counter* (or *servo*), *successive approximation* and

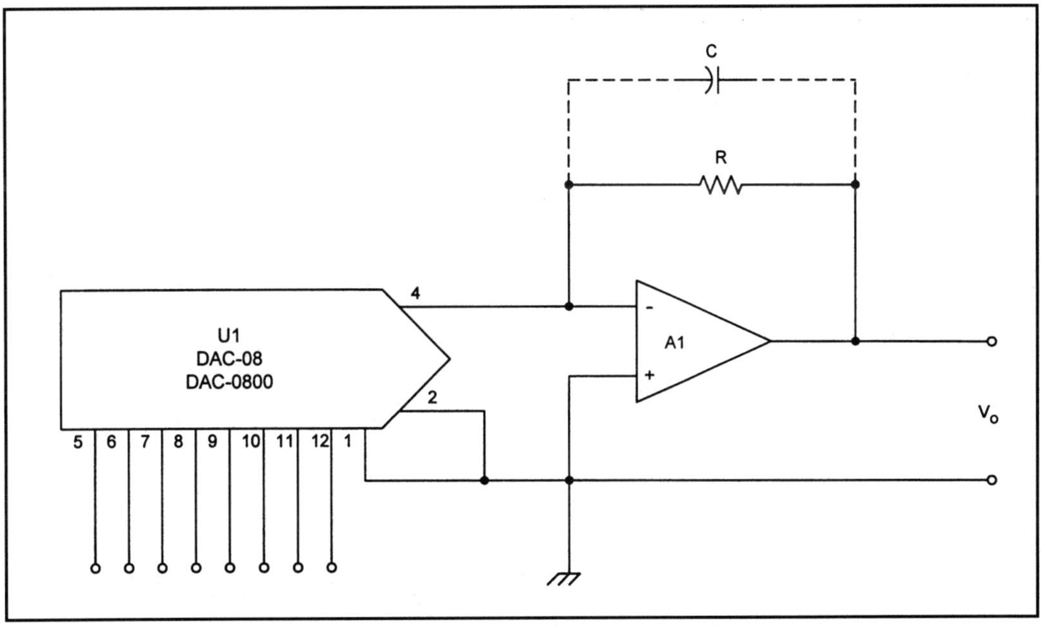

Figure 17-7*. 0-10 volt, TTL compatible circuit with low-impedance voltage output.*

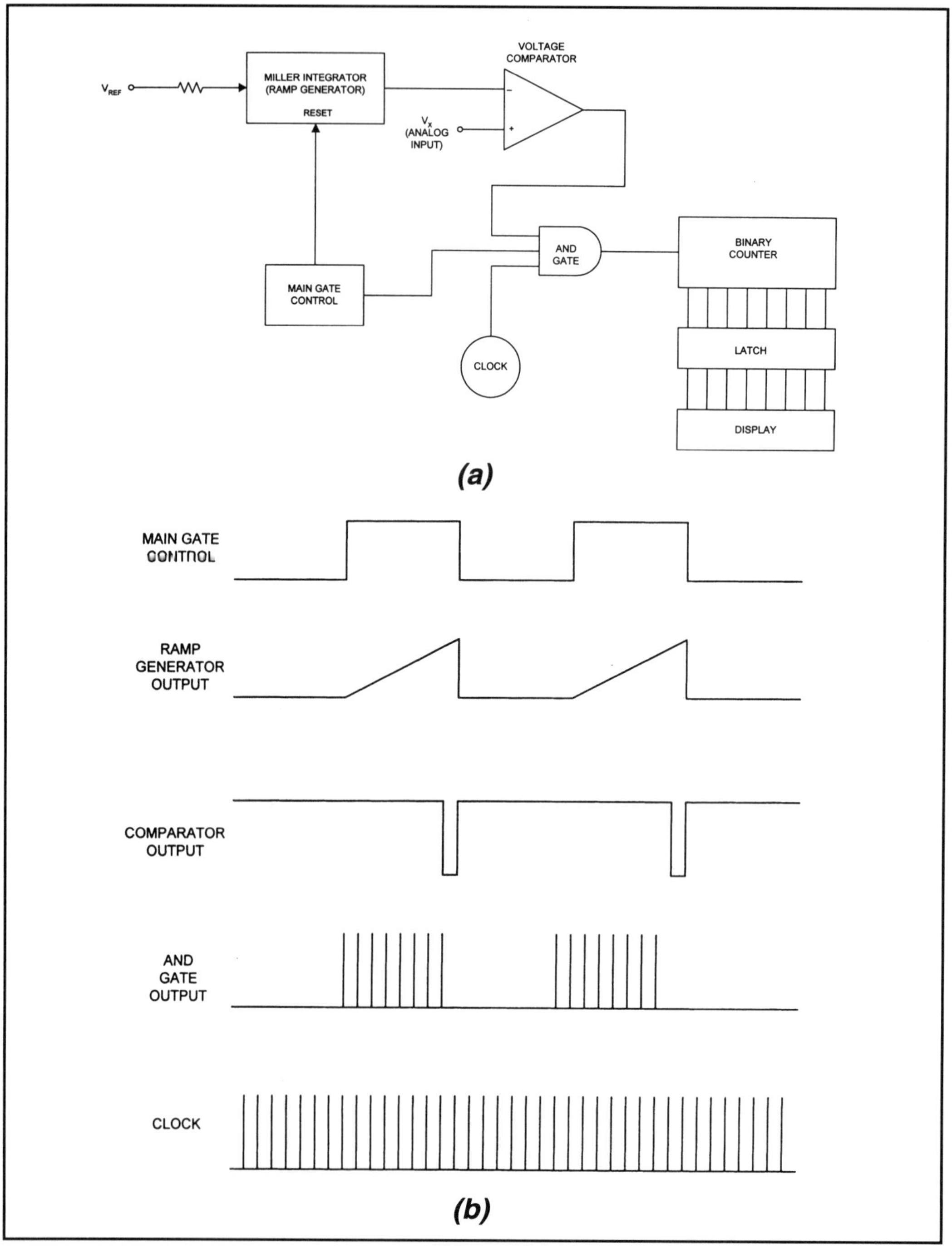

***Figure 17-8**. a. Single-slope A/D converter; b. timing diagram.*

flash methods. The basic size of circuit that we will use is the eight-bit A/D converter, which for many purposes is all that we need. These same discussions are also useful for 10-bit, 12-bit, or higher-order A/D converters.

Integration A/D Methods

Most digital panel meters (DPM) and digital multimeters (DMM) use either the *single integration* or *dual-slope integration* methods for the A/D conversion process. An example of a single-slope integrator A/D converter is shown in *Figure 17-8a*, while its timing diagram is shown in *Figure 17-8b*. The single-slope integrator is simple, but is limited to applications that can tolerate accuracy of one or two percent.

The single-slope integrator A/D converter of *Figure 17-8a* consists of five basic sections: *ramp generator*, *comparator*, *logic*, *clock* and an *output encoder* consisting of a binary counter, latch and display. The ramp generator is an ordinary operational amplifier Miller integrator with its input connected to a stable fixed reference voltage source. This makes input current I_{ref} essentially constant; so the voltage at point B will rise in a nearly linear manner, creating the voltage ramp.

The comparator is an operational amplifier that has an open feedback loop. The circuit gain is the open-loop gain (A_{vol}) of the device selected — typically very high even in low-cost operational amplifiers. When analog input voltage V_x is greater than the ramp voltage, the output of the comparator is saturated at a logic-HIGH level.

The logic section consists of a main AND gate, a main-gate generator, and a clock. The waveforms associated with this circuit are shown in *Figure 17-8b*. When the output of the main-gate control logic section is LOW, switch S1 remains closed, so the ramp voltage is zero. The main-gate signal at point-A is a low-frequency square wave with a frequency equal to the desired time-sampling rate. When point-A is HIGH, S1 is open, so the ramp will begin to rise linearly. When the ramp voltage is equal to the unknown input voltage, V_x, the differential voltage seen by the comparator is zero; so its output drops LOW.

The AND gate requires all three inputs to be HIGH before its output can be HIGH also. The output of the AND gate will go HIGH every time the clock signal is also HIGH.

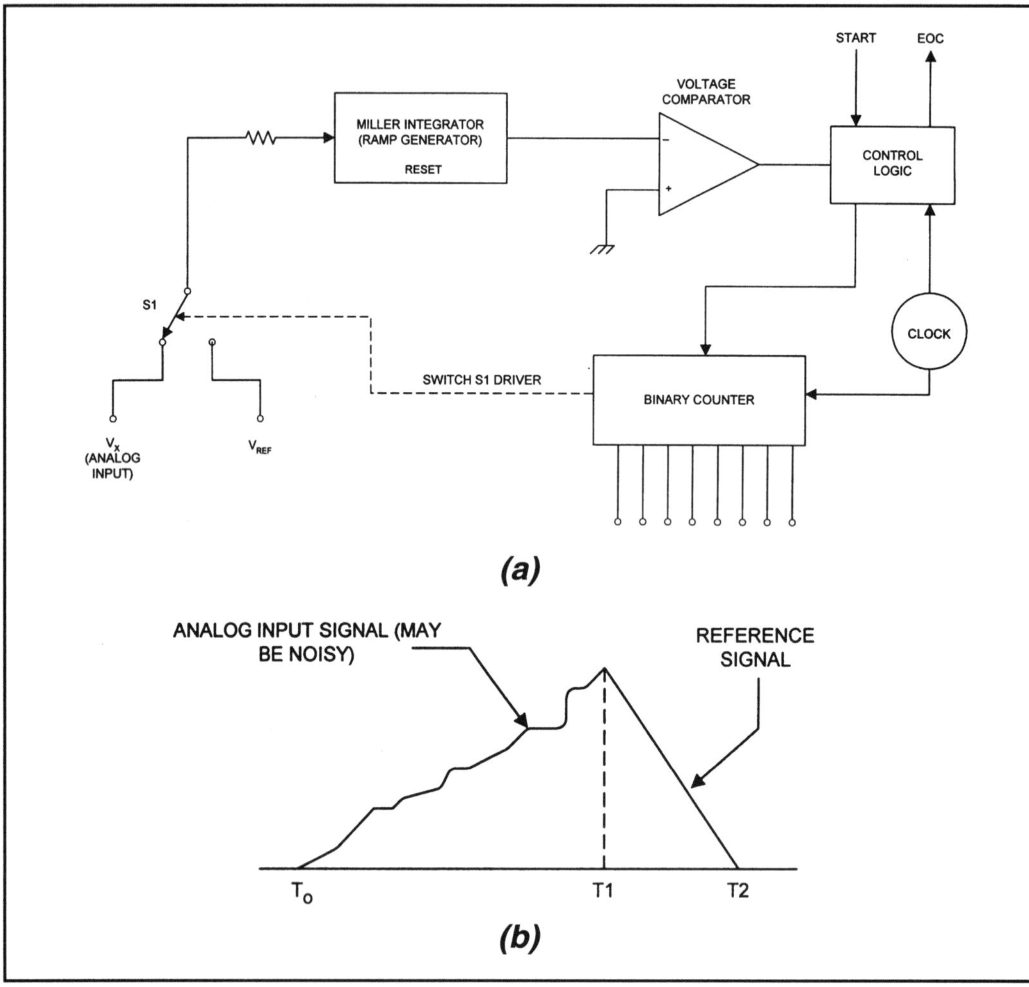

***Figure 17-9**. a. Dual-slope A/D converter; b. timing diagram.*

The encoder (in this case, an eight-bit binary counter) will then see a pulse train with a length proportional to the amplitude of the analog input voltage. If the A/D converter is designed correctly, then the maximum count of the encoder will be proportional to the maximum range (full-scale) value of V_x.

Several problems are found in a single-slope integrator A/D converter:

1. The ramp voltage may be nonlinear.
2. The ramp voltage may have too steep or too shallow a slope.

3. The clock pulse frequency could be wrong.
4. It may be prone to changes in apparent value of V_x caused by noise.

Many of these problems are corrected by the dual-slope integrator of *Figure 17-9a* (timing circuit in *Figure 17-9b*). This circuit also consists of five basic sections: *integrator*, *comparator*, *control logic section*, *binary counter*, and a *reference current* or *voltage source*. An integrator is made with an operational amplifier connected with a capacitor in the negative feedback loop, as was the case in the single-slope version. The comparator in this circuit is also the same sort of circuit as was used in the previous example. In this case, though, the comparator is ground-referenced by connecting +IN to ground.

When a *start* command is received, the control circuit resets the counter to 00000000_2, resets the integrator to 0 volts (by discharging C1 through switch S1), and sets electronic switch S2 to the analog input (position A). The analog voltage creates an input current to the integrator which causes the integrator output to begin charging capacitor C1; the output voltage of the integrator will begin to rise. As soon as this voltage rises a few millivolts above ground potential (0VDC), the comparator output snaps HIGH-positive. A HIGH comparator output causes the control circuit to enable the counter, which begins to count pulses.

The counter is allowed to overflow, and this output bit sets switch S2 to the reference source (position B). The graph of *Figure 17-9b* shows the integrator charging during the interval between *start* and the overflow of the binary counter. At time T2, the switch changes the integrator input from the analog signal to a precision reference source. Meanwhile, at time T2 the counter had overflowed, and again it has an output of 00000000_2 (maximum counter + 1 more count is the same as the initial condition). It will, however, continue to increment so long as we have a HIGH comparator output. The charge accumulated on capacitor C1 during the first time interval is proportional to the average value of the analog signal that existed between T1 and T2.

Capacitor C1 is discharged during the next time interval (T2-T3). When C1 is fully discharged, the comparator will see a ground condition at its active input; so it will change state and make its output LOW. Even though this causes the control logic to stop the binary counter, it does not reset the binary counter. The binary word at the counter output at the instant it is stopped is proportional to the average value of the analog waveform over the interval (T1-T2). An *end-of-conversion* (EOC) signal is generated to notify the computer that the output data is both stable and valid (and therefore ready for use).

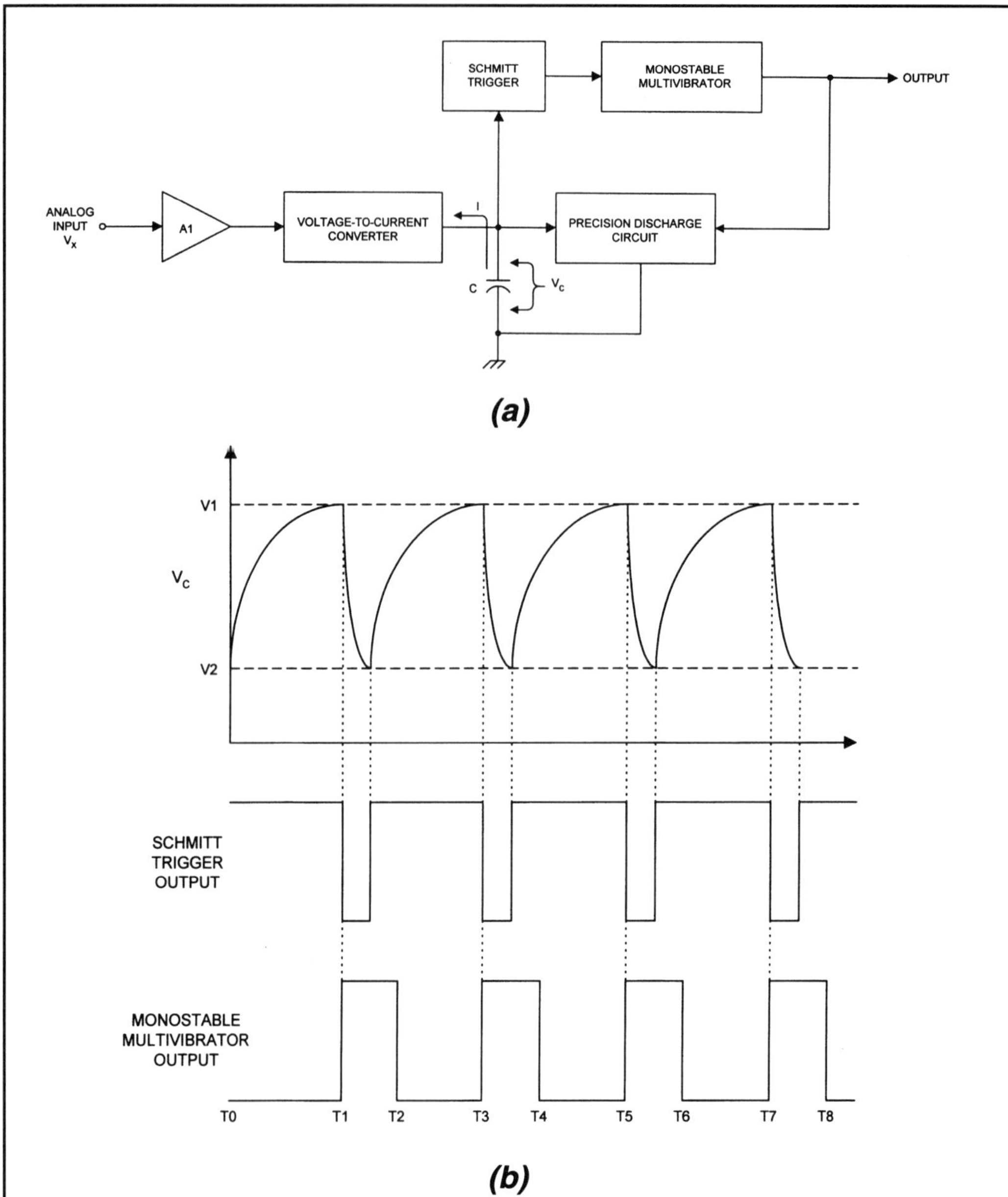

***Figure 17-10**. a. Voltage-to-frequency converter; b. timing diagram.*

Voltage-to-Frequency Converters

These circuits are not A/D converters in the strict sense, but are very good for representing analog data in a form that can be taped on a low-cost audio machine, or transmitted over radio. The V/F converter output can also be used for direct input to a computer if a binary counter is used to measure the output frequency. Two forms of V/F converter are common. One is a *voltage-controlled oscillator* (VCO); that is, a regular oscillator circuit in which the output frequency is a function of an input control voltage. If the VCO is connected to a binary or binary coded decimal (BCD) counter, then the VCO becomes a V/F form of A/D converter.

The type of V/F converter shown in *Figure 17-10* is superior to the VCO method. The circuit is shown in *Figure 17-10a*, while the timing waveforms are shown in *Figure 17-10b*. The operation of this circuit is dependent upon the charging of a capacitor, although not an RC network, as in the case of some other oscillator or timer circuits. The input voltage signal (V_x) is amplified (if necessary) by A1, and then converted to a proportional current level in a *voltage-to-current converter* stage. If the voltage applied to the input remains constant, so will the current output of the V-to-I converter (I).

The current from the V-to-I converter is used to charge the timing capacitor, C. The voltage appearing across this capacitor (V_c) varies with time as the capacitor charges (see the V_c waveform in *Figure 17-10b*). The precision discharge circuit is designed to discharge capacitor C to a certain level (V2) whenever the voltage across the capacitor reaches a predetermined value (V1). When the voltage across the capacitor reaches V2, a Schmitt trigger circuit is fired, turning on the precision discharge circuit. The precision discharge circuit, in its turn, will cause the capacitor to discharge rapidly but in a controlled manner to value V1. The output pulse snaps HIGH when the Schmitt trigger fires (or the instant V_c reaches V1), and drops LOW again when the value of V_c has discharged to V2. The result is a train of output pulses whose repetition rate is exactly dependent upon the capacitor charging current which, in turn, is dependent upon the applied voltage. Hence, the circuit is a voltage to frequency converter.

Like the VCO circuit, the output of the V/F converter can be applied to the input of a binary counter. The parallel binary outputs become the data lines to the computer. Alternatively, if the frequency is relatively low, the computer can be programmed to measure the period between pulses. Also, certain computer interface chips have built-in timers that can measure the period.

Counter Type (Servo) A/D Converters

A counter type A/D converter (also called "servo" or "ramp" A/D converters) is shown in *Figure 17-11*. It consists of a *comparator*, *voltage output DAC*, *binary counter* and the necessary *control logic*. When the *start* command is received, the control logic resets the binary counter to 00000000_2, enables the clock, and begins counting. The counter outputs control the DAC inputs; so the DAC output voltage will begin to rise when the counter begins to increment. As long as analog input voltage V_{in} is less than V_{ref} (the DAC output), the comparator output is HIGH. When

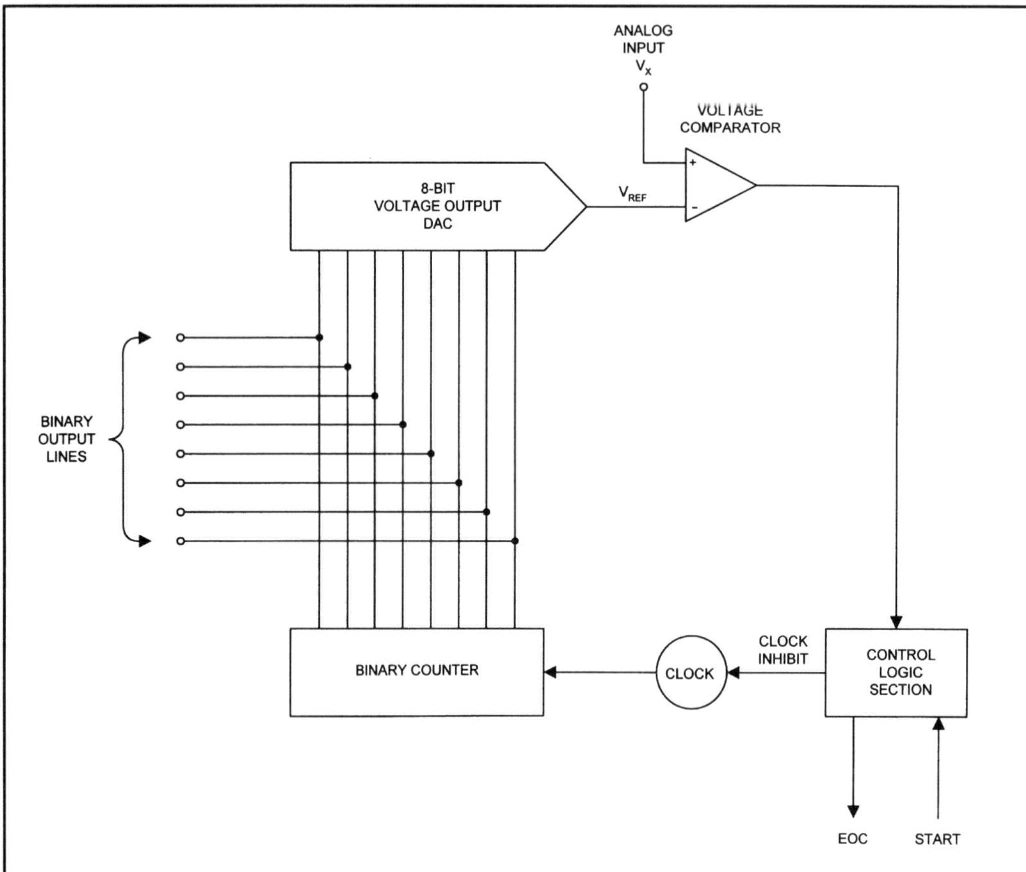

***Figure 17-11**. Binary or "servo" A/D converter.*

V_{in} and V_{ref} are equal, however, the comparator output goes LOW, which turns off the clock and stops the counter. The digital word appearing on the counter output at this time represents the value of V_{in}.

Both slope and counter type A/D converters take too long for many applications, on the order of 2^N clock cycles (where N = number of bits). Conversion time becomes critical if a high frequency component of the input waveform is to be faithfully reproduced. Nyquist's criteria requires that the sampling rate (or conversions per second) be at least twice that of the highest Fourier frequency to be recognized.

Successive Approximation A/D Converters

Successive approximation A/D conversion is best suited for many applications where speed is important. This type of A/D converter requires only N + 1 clock cycles to make the conversion, and some designs allow truncation of the conversion process after fewer cycles if the final value is found prior to N + 1 cycles.

The successive approximation converter operates by making several successive trials at comparing the analog input voltage with a reference generated by a DAC. An example is shown in *Figure 17-12*. This circuit consists of a *comparator*, *control logic section*, a *digital shift register*, *output latches* and a *voltage output DAC*.

When a START command is received, a binary 1 (HIGH) is loaded into the MSB of the shift register, which sets the output of the MSB latch HIGH. A HIGH in the MSB of a DAC will set output voltage V_{ref} to half-scale. If input voltage V_{in} is greater than V_{ref}, the comparator output stays HIGH, and the HIGH in the shift register MSB position shifts one bit to the right; therefore, it occupies the next most significant bit (bit 2). Again the comparator compares V_{in} with V_{ref}. If the reference voltage from the DAC is still less than the analog input voltage, the process will be repeated with successively less significant bits until either a voltage is found that is equal to V_{in} (in which case the comparator output drops LOW) or the shift register overflows.

If, on the other hand, the first trial with the MSB indicates that V_{in} is less than the half-scale value of V_{ref}, the circuit continues making trials below V_{ref}. The MSB latch is reset to LOW, and the HIGH in the MSB shift register position shifts one-bit to the right to the next most significant bit (bit 2). Here the trial is repeated again.

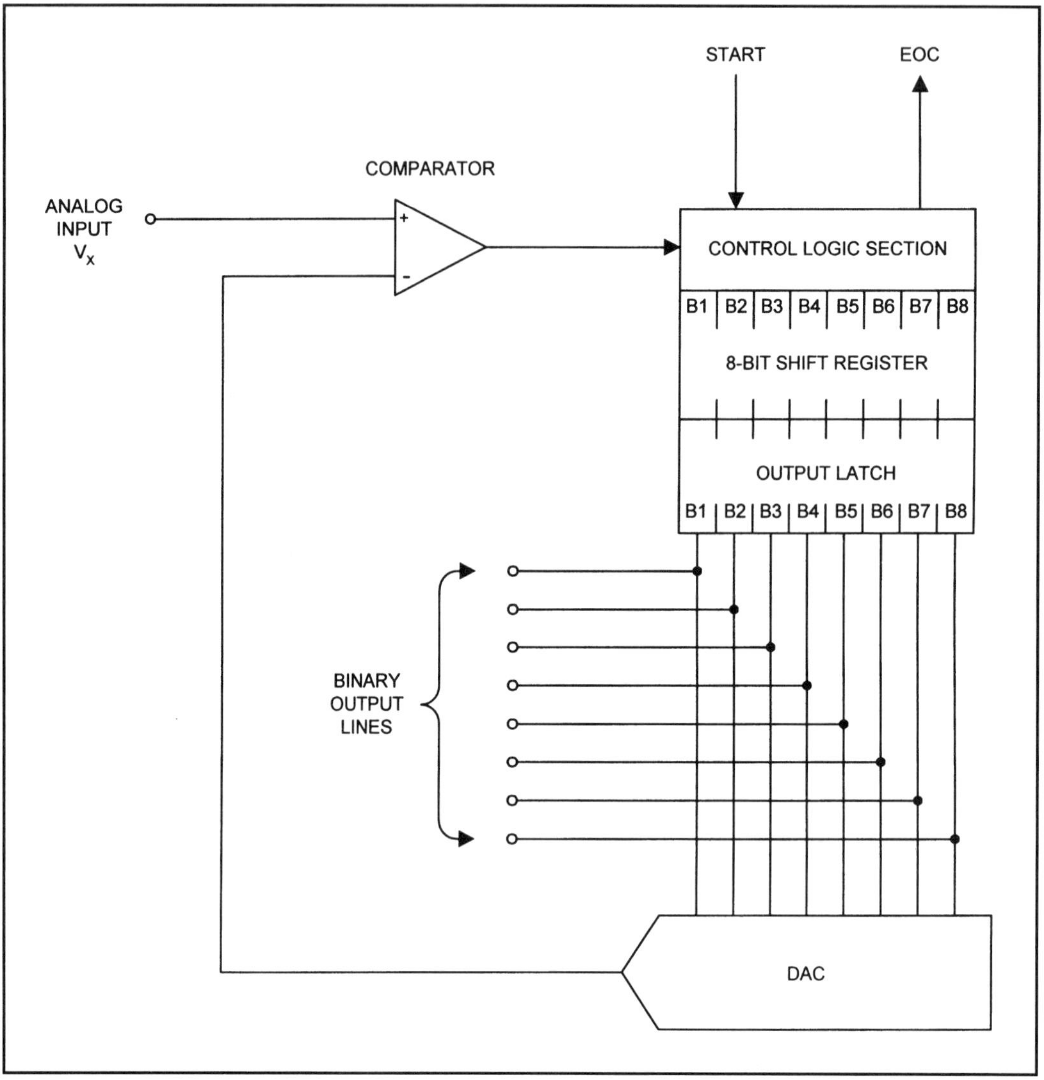

Figure 17-12*. Successive approximation register A/D converter.*

This process will continue as before until either the correct level is found, or overflow occurs. At the end of the last trial (bit 8 in this case), the shift register overflows and the overflow bit becomes an end-of-conversion (EOC) flag to tell the rest of the world that the conversion is completed.

This and most other types of A/D converters require a starting pulse and signal completion with an EOC pulse. This requires the computer or other digital instrument to

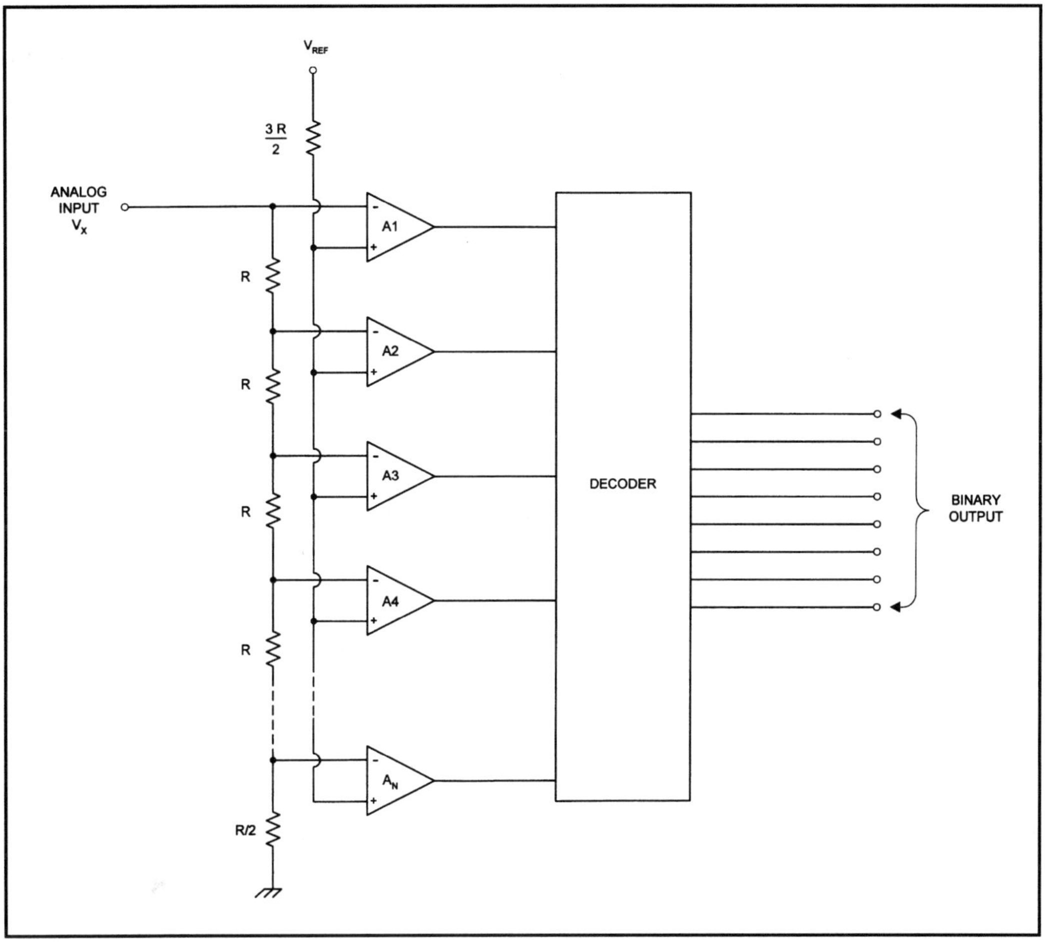

Figure 17-13. *Flash or parallel A/D converter.*

engage in "bookkeeping" to repeatedly send the *start* command and look for the EOC pulse. If the *start* input is tied to the EOC output, then conversion is continuous; the computer need only look for the periodic raising of the EOC flag to know when a new conversion is ready. Such operation is said to be *asynchronous*.

Parallel or "Flash" A/D Converters

The parallel A/D converter (*Figure 17-13*) is probably the fastest A/D circuit known. Indeed, the fastest ordinary commercial products use this method. Some sources call the parallel A/D converter the "flash" circuit because of its inherent high speed.

The parallel A/D converter consists of a bank of $[2^N - 1]$ voltage comparators biased by reference potential V_{ref} through a resistor network that keeps the individual comparators 1-LSB apart. Since the input voltage is applied to all the comparators simultaneously, the speed of conversion is limited (essentially) by the slew rate of the slowest comparator in the bank, and also by the decoder circuit propagation time. The decoder converts the output code to the binary code needed by the computers.

The analog-to-digital converter (ADC or A/D) is a circuit that is used to produce a *binary* (or base 2) *number output* that represents an *analog voltage* applied to the input (*Figure 17-14*). It is the inverse process to the digital-to-analog converter (DAC), which produces an analog output voltage that is proportional to the binary number applied to the input.

The MAX-187 12-Bit Serial Output A/D Chip

A company called *Maxim Integrated Products* (120 San Gabriel Drive, Sunnyvale, CA 94086) produces a number of clever products whose designs suggest their product design engineers have "been there, seen that, done it, bought the tee-shirt," then went back home to California to design products that real engineers can use. I first became aware of Maxim products when I used some of their RS-232 interface products in a project. The product that concerns us here is one of the many A/D converter chips made by Maxim.

One reason for choosing the Maxim MAX-187 device is that you can easily buy this chip for about $16 from *Digi-Key* (P.O. Box 677, Brooks Avenue South, Thief River

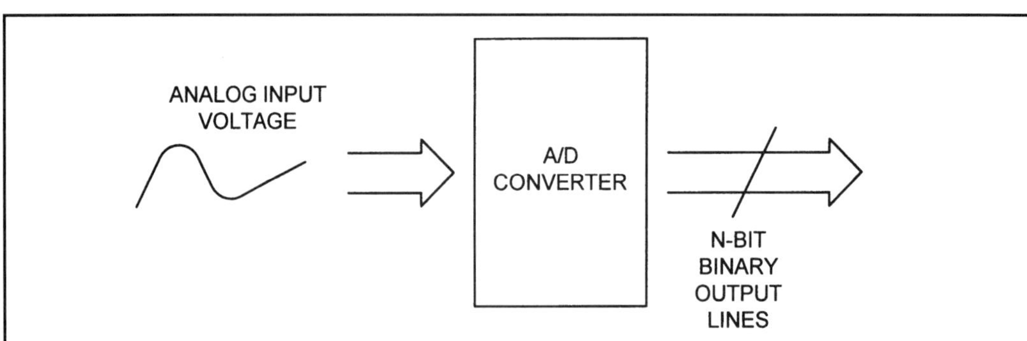

Figure 17-14. *Basic function of the A/D converter is to provide a binary output that is proportional to the applied analog voltage.*

Falls, MN 56701-0677; 1-800-344-4539). A lot of companies make interesting A/D converter products, but do not have distributors who are willing to sell unit quantities to hobbyists (or others who don't need a gazillion of them).

Another reason for selecting the MAX-187 is that it is available in the 8-pin miniDIP package, as well as the surface mount SO packages. Most hobbyists will want to use the miniDIP package.

The MAX-187 (*Figure 17-15*) is a serial output, 12-bit A/D converter with an input voltage range of 0 to +5 volts. Because it is a 12-bit device, it will break the 5V input range into 212 = 4,096 different states according to the code:

Voltage	**Binary Code**
0 volts	000000000000_2
2.5 volts	100000000000_2
4.9988 volts	111111111111_2

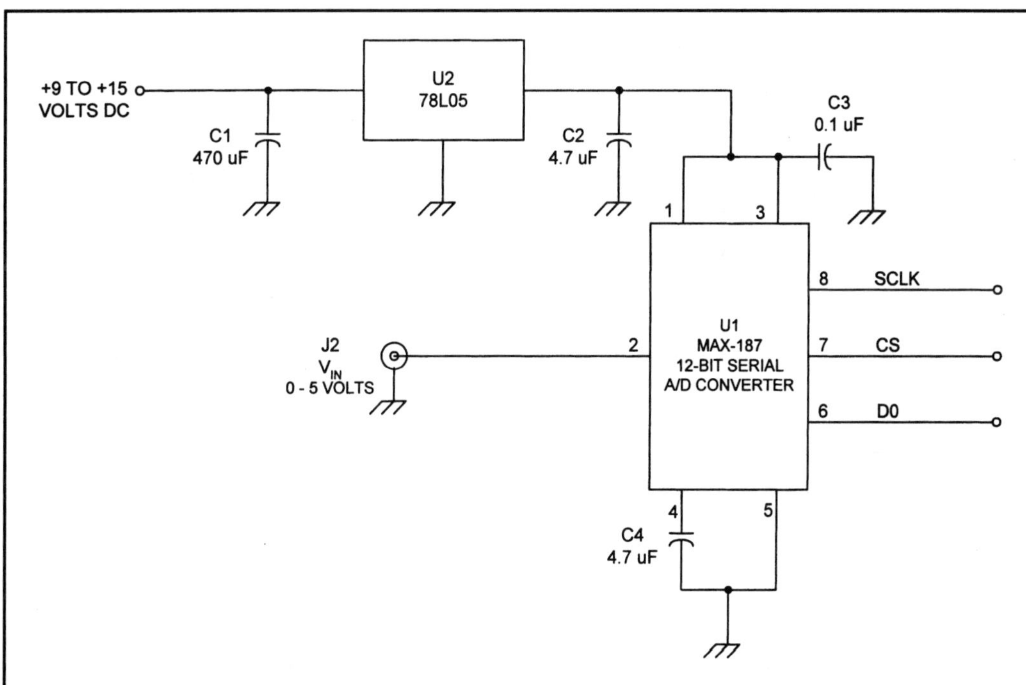

***Figure 17-15**. Basic circuit for the MAX-187. The 78L05 is a +5V voltage regulator capable of 100 mA operation.*

Of course, values between these three points are represented with other binary numbers proportional to the actual voltage. The MAX-187 contains its own internal track-and-hold circuit that does not need any external holding capacitor. It also has its own internal 4.096V buffered reference source (that is, V_{ref} is internal to the chip). The sampling rate is 75 kHz.

The pinouts for the 8-pin miniDIP version of the MAX-187 are detailed in *Figure 17-16*. These pins serve the following functions:

Pin Number 1, V_{dd}. DC power. This pin receives the V+ voltage, which must be between +4.75 and +5.25 volts.

***Pin Number 2, AIN*.** Analog input. This pin receives the analog input voltage to be converted. The voltage must be kept within the range -0.3 volts and (V_{dd} + 0.3) volts.

***Pin Number 3, SHDN (Active LOW)*.** Shut down. This terminal can be used for providing a power saving shutdown mode. When SHDN is low (or at zero volts

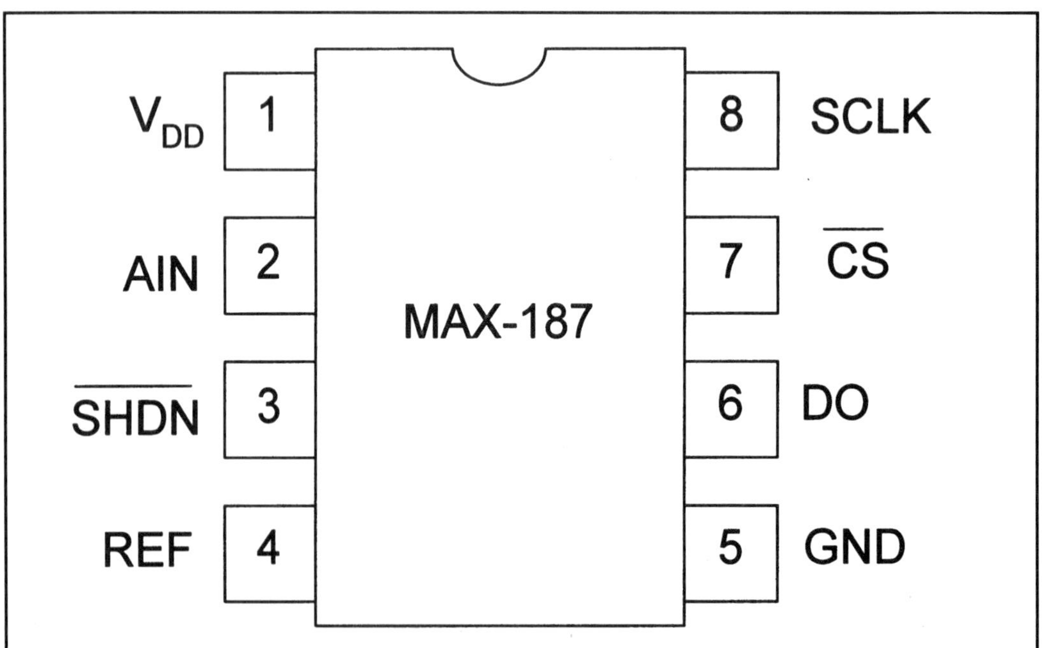

***Figure 17-16**. Pinouts of the MAX-187.*

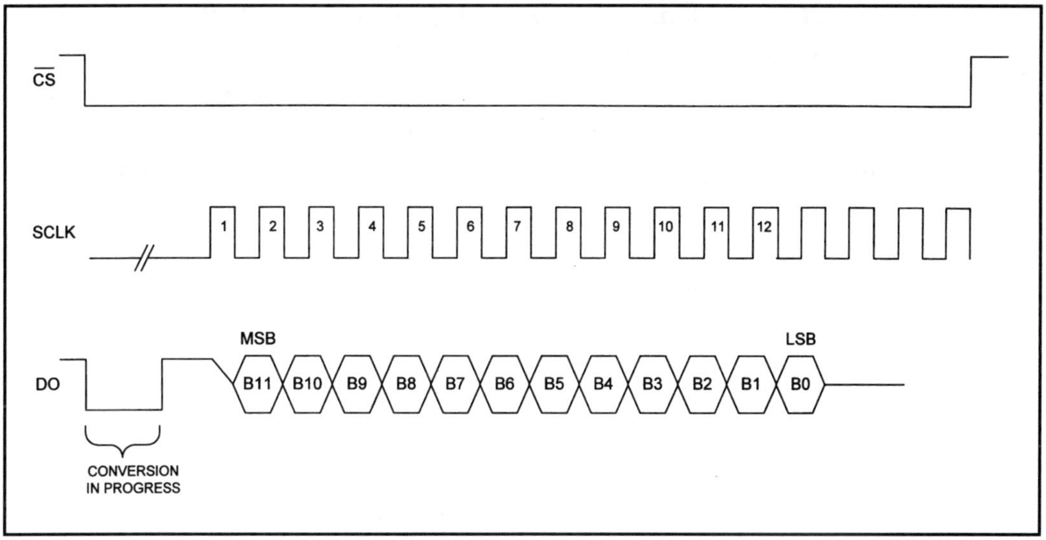

***Figure 17-17**. Timing diagram of the MAX-187.*

[grounded]), the MAX-187 is in the shutdown mode and consumes only 10 µW. When SHDN is tied HIGH (as in when it is connected to V_{dd}), the MAX-187 is enabled and draws about 7.5 mW. The HIGH condition also enables the internal reference voltage. If the SHDN is left floating (has no connection to either ground or V_{dd}), then the MAX-187 is enabled but the internal reference voltage is disabled.

Pin Number 4 REF. Reference voltage. When SHDN is HIGH, then this terminal must be bypassed to ground with a 4.7 µF capacitor (internal reference operation). When SHDN is left floating, then the external reference voltage (+2.5 to + V_{dd}) must be applied.

Pin Number 5 GND. Ground. Connects to the signal and power supply common line.

Pin Number 6 DO. Data out. Serial data output line. Data changes on the falling edge of SCLK.

Pin Number 7 CS (Active LOW). Chip select. Pulling this pin LOW selects the chip and initiates the conversion on its falling edge. Setting CS to HIGH deselects the chip and puts the data output line (DO) into a high impedance state.

Pin Number 8 SCLK. System clock. This clock is used to clock data out of the successive approximation register at rates up to 5 MHz.

The timing diagram for the MAX-187 is shown in *Figure 17-17*. The timing for the three serial interface lines (CS, SCLK, and DO) are shown. Conversion starts when the chip select line (CS, pin number 7) is brought LOW; so conversion begins on the falling edge. At this time, the data output line (DO, pin number 6) drops LOW. The *end of conversion* (EOC) is signaled by the data output line (DO) going HIGH again.

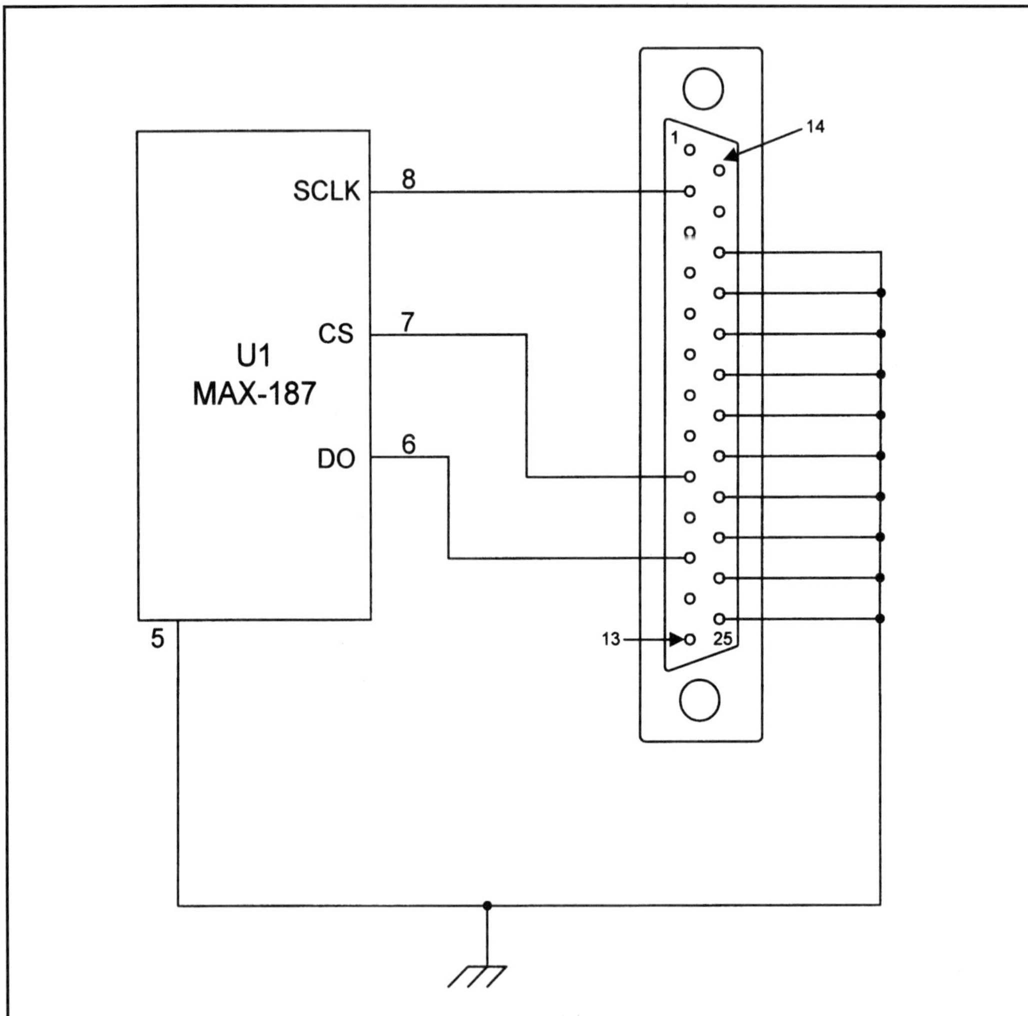

Figure 17-18. *Connection of the MAX-187 to a parallel printer port interface.*

At this point, clock pulses on SCLK (pin number 8) can begin clocking data bits out through the DO line, starting with the bit B11, the *most significant bit* (MSB) and finishing with bit B0, the *least significant bit* (LSB). Any interface software written for this chip must account for these factors.

One possibility for using the MAX-187 is to connect it to your computer through the parallel printer port (LPT1, LPT2, etc). Although the data lines of the printer port are outputs only, there are several handshaking lines that can be used to receive input data.

It's these lines that make all those low-cost parallel printer port add-ons for PCs possible (I have an Iomega ZIP drive, an EZ-Photo scanner and several other devices that operate from my parallel printer port).

The circuit for the parallel printer port interface is shown in *Figure 17-18*. All of the higher order lines of the DB-25 connector, from pin number 16 to pin number 25, are grounded. Three of the remaining pins are used for the interface:

DB-25 Pin Number	**MAX-187 Signal**
2	SCLK
9	CS
11	DO

For those who want to experiment with the MAX-187 device, Maxim makes an evaluation kit at a reasonable price. If you want to spend more money, they also have a controller card available and accompanying software.

Index

SYMBOLS

A

C

D

G

H

P

PROMPT®
PUBLICATIONS

PC Hardware Projects, Volume 1

by James "J.J." Barbarello

As an electronics hobbyist and PC owner, you probably own a digital multimeter and a logic probe. These inexpensive test and measurement instruments provide enough test muscle for many tasks, but just don't have enough capability for serious work in the world of digital electronics.

Using commonly available components and standard construction techniques, *PC Hardware Projects* will guide you through the construction of an 8-In/8-Out channel logic analyzer and a 25-line multipath continuity tester. Once you are done, you will be able to put them to immediate use with the full-featured software included with the book. You will also be able to combine the 8/8CLA, the MCT25, appropriate power supply, and a prototyping solderless breadboard system into a single digital workstation interface. *PC Hardware Projects* guides you through every step of the construction process and shows you how to check your progress.

Projects
256 pages • paperback • 7-3/8 x 9-1/4"
ISBN: 0-7906-1104-X • Sams: 61104
$29.95

PC Hardware Projects, Volume 2

by James "J.J." Barbarello

If you have always wanted to work with stepper motors and need a rig to drill your own PC boards quickly and accurately, get out your soldering iron and follow along. The second volume in a two-part series, *PC Hardware Projects* discusses stepper motors, how they differ from conventional and servo motors, and how to control them. It investigates different methods to control stepper motors, and provides you with circuitry for a dedicated IC controller and a component hardware controller.

Then *PC Hardware Projects* guides you through every step of the construction process of an automated, PC-controlled PCB drilling machine which can drill as many as 500 holes in a 6 x 8" PCB with excellent repeatability and resolution. With construction completed, you'll then walk through an actual design layout, creating a PC design and board. You'll see how the drill data file is determined from the layout, and drill the PCB.

Projects
191 pages • paperback • 7-3/8 x 9-1/4"
ISBN: 0-7906-1109-0 • Sams: 61109
$29.95

CALL 1-800-428-7267 TODAY FOR THE NAME OF YOUR NEAREST PROMPT PUBLICATIONS DISTRIBUTOR

Prices subject to change.